Dublin Gulch

DUBLIN GULCH

A History of the Eagle Gold Mine

Michael Gates

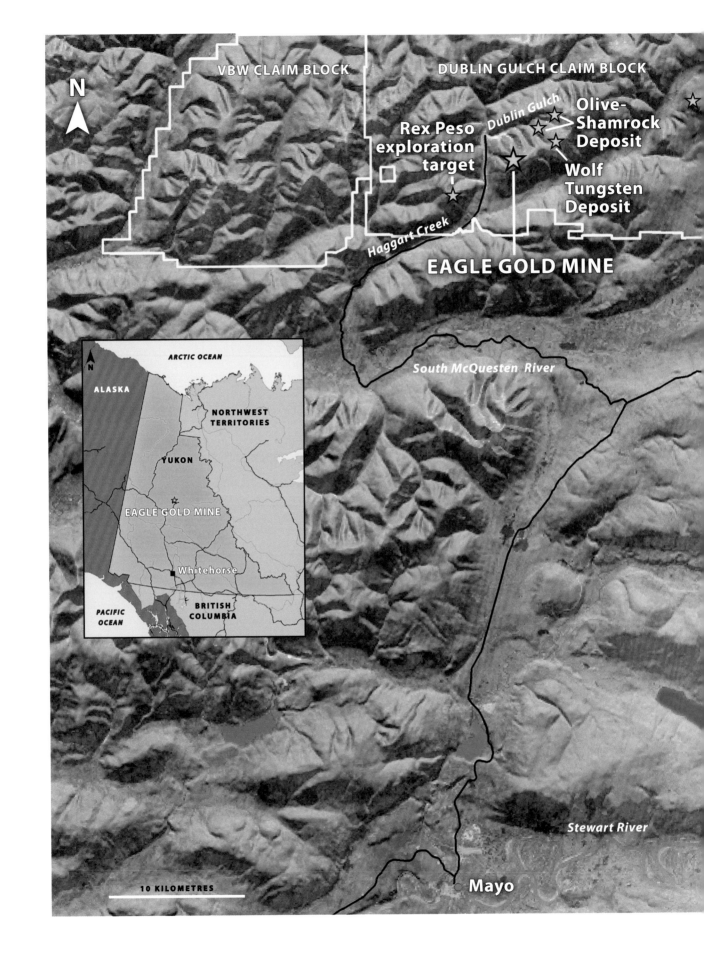

VBW CLAIM BLOCK

DUBLIN GULCH CLAIM BLOCK

N

Dublin Gulch

Olive-
Shamrock
Deposit

Rex Peso
exploration
target

Wolf
Tungsten
Deposit

Haggart Creek

EAGLE GOLD MINE

South McQuesten River

ARCTIC OCEAN

N

ALASKA

NORTHWEST
TERRITORIES

YUKON

EAGLE GOLD MINE

Whitehorse

PACIFIC
OCEAN

BRITISH
COLUMBIA

Stewart River

10 KILOMETRES

Mayo

Nugget
exploration
target

Elsa

Keno City

Mayo Lake

N

WASTE ROCK
STORAGE

CONVEYOR

HEAP LEACH
FACILITY

Dublin Gulch

GOLD RECOVERY
PLANT

CAMP

500 METRES

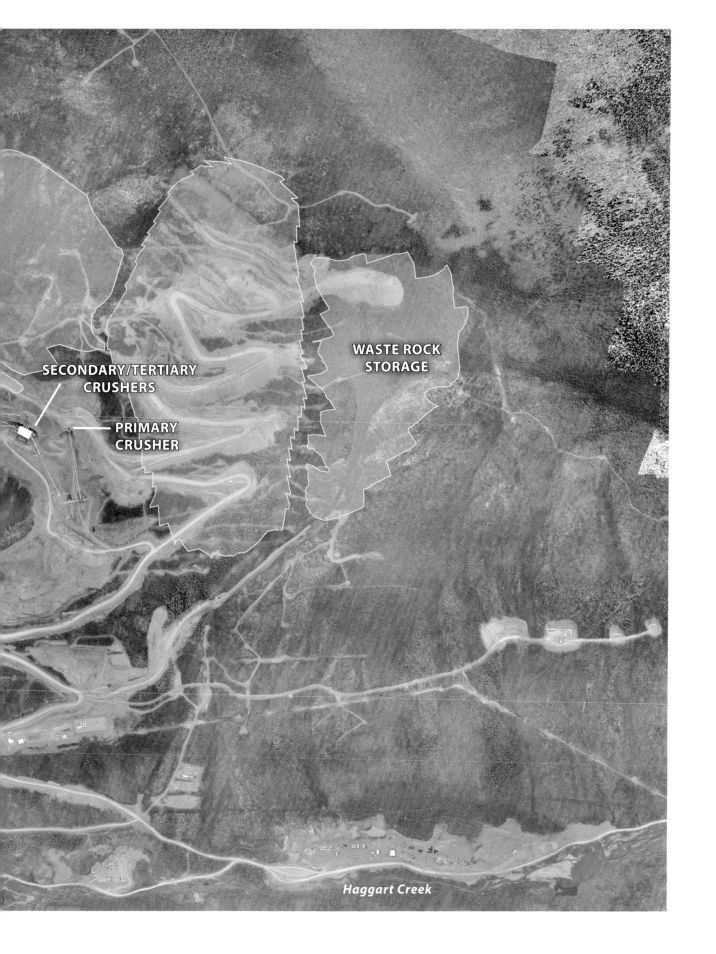

SECONDARY/TERTIARY
CRUSHERS

PRIMARY
CRUSHER

WASTE ROCK
STORAGE

Haggart Creek

Contents

FOREWORD by John McConnell ▬ xi

FOREWORD by Sandy Silver ▬ xiii

FOREWORD by Simon Mervyn ▬ xv

Preface ▬ xvii

CHAPTER 1 **Beginnings** ▬ 1

CHAPTER 2 **Placer** ▬ 7

CHAPTER 3 **Hardrock** ▬ 45

CHAPTER 4 **Growing Pains** ▬ 75

CHAPTER 5 **The Eagle Is Born** ▬ 119

CHAPTER 6 **Conclusion** ▬ 171

ENDNOTES ▬ 187

GLOSSARY ▬ 197

SELECTED SOURCES ▬ 205

ACKNOWLEDGEMENTS ▬ 209

IMAGE CREDITS ▬ 211

INDEX ▬ 213

ABOUT THE AUTHOR ▬ 219

Foreword

John McConnell, President and CEO of Victoria Gold

For over ten years, my focus has been on seeing the Eagle Gold Mine through all the phases of development, from exploration to operation. I am proud that we have achieved this vision; many mines never come to fruition for technical, social or economic reasons. But I am an optimist, which has carried me through the highs and lows with the Eagle Project.

Over my career working in northern Canada, I have gained an appreciation for this vast land, the environment and the protection of it, First Nations and others who live here. It is my core belief that any development must be done with respect. Any development will change the way things are, but it must be done responsibly and in a way that improves the lives of those it affects.

We have worked with the Na-Cho Nyäk Dun First Nation, in whose traditional territory Eagle is located, to build a lasting partnership and long-term benefits. Chief Mervyn has been the chief for the majority of the process, and he has been a leader, teacher and friend. Thank you, Chief Mervyn, for your perseverance and dedication to your people.

As of January 1, 2020, more than two million work-hours have passed since the last lost-time incident. This is unprecedented in construction and mining, and is something we are all very proud of. Our "Yukoners, It's Time to Come Home" campaign has also been very successful. More than half our workforce are Yukoners, and about 25 percent are Indigenous. Also, 25 percent of our workforce are women. Some of the stories of the people who work for Victoria Gold bring tears to my eyes—and will form the next collection of stories told of Dublin Gulch.

Of course, I didn't do this alone. If it were not for our team of professionals, many of whom have become personal friends over the last decade, as well as my board of directors, government regulators, First Nations representatives and many other individuals in the community, we would never have reached the point we are at today.

While most of my career has been working in Canada's north, I am a relative new-comer to the Yukon. During these ten years, I met my wife, Tara, and established a home in Whitehorse and another closer to work in Mayo. Our daughter, Katie, was born in the Yukon and my daughter, Lauren, was born and grew up in the north. We want the Yukon to be the kind of place for everyone to live and raise a family. I could not have overcome the numerous hurdles of this journey without the support of Tara, her parents Jim and Dagmar Christie, and her brother Sheamus Christie and his wife Kara Christie, who have helped us on the home front when our lives were hectic. Also, a big thank you to executive affairs manager Lenora Hobbis, who has assisted me for over twenty years, the last ten at Victoria Gold, including helping to pull together the photos and finalizing this book.

Dublin Gulch has been the focus of interest for generations of prospectors and miners for more than a century. It is that tradition of enterprise and commitment that has, in part, brought us to this day. We have succeeded in bringing into production the largest gold mine in Yukon history. We look forward to many years of prosperity and opportunity for those who call Yukon their home. This book is the story of Dublin Gulch and the long road to the opening of the Eagle Gold Mine. I hope you enjoy it.

Foreword

Sandy Silver, Yukon Premier

The Dublin Gulch Eagle Project's impact on Yukon's economy is tremendous. In the many stages of development—before a single ounce of gold was poured—$150 million for contracting and services went straight into the pockets of Yukon businesses. Almost a million hours of labour were soaked up by Yukoners. Looking ahead, the Eagle Project will give $110 million in royalties to Yukon, and another $240 million to this territory in income tax. This is the largest gold mine in the notable history of mining in Yukon.

A more remarkable consideration, one of national importance, is how the Dublin Gulch project has come to epitomize a Yukon way of mining. By modelling a good relationship with the First Nation of Na-Cho Nyäk Dun, and reducing the project's carbon footprint, Victoria Gold provides a world-class standard for responsible mining. Take, for example, the company's decision to connect to our hydro-electrical grid in 2017. This investment not only reduces the emissions of the mine, but the company's investment will also generate an additional ten million dollars of revenue a year for Yukon Energy over the ten-year life of the project, and will benefit Yukoners by decreasing power rates.

Just as evidence of gold deposits is essential to the success of a mine, so too are the reputations and social values of the company's executives and managers. Victoria Gold is not only the largest corporate employer in Yukon, but is also described as "family" by Chief Mervyn of the First Nation of Na-Cho Nyäk Dun. This is no coincidence.

This company has invested heavily in educational initiatives. Through the Every Student, Every Day program, students across our territory arrive to school each and every day ready, willing and able to learn. Victoria Gold successfully advocated for the return of skilled labourers working out of territory with the "Yukoners, It's Time to Come Home" campaign. I have been on site in Mayo enough times to hear many compelling stories of proud Yukoners who have done just that.

I could go on about successful initiatives to maximize First Nations hires, the impressive number of women working on-site, charitable sponsorships and donations, but I digress. All this to say it took an exceptional and progressive team to get the Eagle Gold project to production. With intestinal fortitude and enormous impact, John McConnell and Victoria Gold are rewriting Yukon's mining history. This book is the story of David in a world of Goliaths. Here is a small company that put in a Herculean effort to get the Dublin Gulch project off the ground, a company whose primary capital was sharp minds and entrepreneurial spirit. Its success is as much of an industry anomaly as gold is a geophysical one.

On the surface, it was a series of significant decisions that ushered this project on its way to fruition, but behind every one of those decisions was the company's virtue of making sure it would add value to the lives of the people involved. This book is about what happens when you pair mineral-rich ground with human ingenuity. It is about the inventiveness it took to work around a complex investment climate and unkind capital markets to responsibly secure the equity and loans required to breathe life into a mine. Described as an "emotional roller coaster," by one Victoria Gold employee, this is more than a history of subarctic rocks, dirt and minerals—it is a riveting story of people making their dreams come true.

Foreword

Simon Mervyn, Chief of the First Nation of Na-Cho Nyäk Dun

When I was young, I worked for Klaus Djukastein on a mining operation not far from the present location of the Eagle Mine on Dublin Gulch. It was a simple operation consisting of a D7 bulldozer pushing gravel into a sluice box, moving the tailings and building a tailings pond. In the fall of 2019, looking over the same scene where the Eagle Mine now operates, I saw a complex arrangement of mechanical systems operated by computer. The mine moves as much ore in a few days as we did during an entire summer. What a long way we have come!

It was not an easy transition. In fact, it took years of careful consultation and negotiation between Victoria Gold and the First Nation of Na-Cho Nyäk Dun to work out a solid social licence between our two parties. We didn't always agree, but we were patient and eventually ironed out the details that worked for both sides. We have worked with Victoria for many years establishing a mutually beneficial relationship, to the point where Victoria Gold now exemplifies the model for companies wishing to work in our traditional territory.

The First Nations people have used this area for millennia for traditional harvesting; however, we recognize that there are economic benefits in modern times that we can choose to participate in. The First Nation seeks a balance between the protection of the environment, which is the foundation upon which our lives have been built, and economic opportunities as we move together into the future. The Na-Cho Nyäk Dun looks forward to maintaining a strong and lasting relationship with Victoria Gold that will be beneficial for both parties for many years to come.

Preface

When I was first asked to write a book about the history of Dublin Gulch and the Eagle Mine, I wondered: Where would I find enough information to write a decent account of a tiny creek I had barely heard of and couldn't locate on a map? I asked how long it should be. Thirty-five thousand words was the reply. I fussed over that for a while because I wasn't certain I could find enough words to meet the target.

I need not have worried. As I dug into the historical records and interviewed the people who have been involved with Dublin Gulch, my worry became, *How do I keep it down to thirty-five thousand words?* As the story unfolded, there was enough material to fill a large volume. The challenge became one of telling the story without drowning the reader in too much detail.

The evolution of the Eagle deposit into the Yukon's largest gold mine was a complex undertaking that started to be planned in the 1990s. There were stops and starts for many years, and it was not until Victoria Gold acquired the property that the mine eventually became a reality. There were ups and downs along the way. Some events spelled good fortune for Victoria; the financial downturn if 2008 was a stroke of luck for Victoria, for example, while the decline of the price of gold a few years later threatened to put the project to bed for good.

The development of this project by a junior mining company is a remarkable story of the vision, grit and determination of a committed team. I have come to think of it as "the little mine that could," but then, it's not such a little mine. It has become the largest gold mine ever to operate in the Yukon, and it may enrich the economy of the Yukon for many years to come.

In putting this story together, I encountered many differing units of measure. Grams per tonne versus ounces per cubic yard, and miles versus kilometres. I discussed this with Tara Christie, my contact person for the project, and we decided that rather than attempt to convert everything into uniform units of measure, which could result in embarrassing miscalculations on my part, I would leave them as I found them in the source documents. In contemporary context, I use measurements in metric (i.e. kilometres rather than miles), but in direct quotes, or in referring to events from many years ago, I use imperial measures.

I envisioned an audience that might include readers who are not familiar with the technical jargon that is associated with the mining industry, so I tried not to become too technical, while at the same time accurately rendering the techniques and processes that I describe. There is a small glossary of terms at the back of the book that may help clarify some of the technical terms that were used in this account. Any direct quotations that are not otherwise cited come from interviews performed by the author, as documented in the bibliography.

The story of the Eagle Mine is an acknowledgement of the importance of mining to the economics of the territory, and the determined and dedicated individuals who have followed their dreams at Dublin Gulch for more than a century. It took determination, ingenuity and plenty of hard work to make a livelihood extracting minerals from the gravels and the bedrock of the region. This account is not the full story of Dublin Gulch, but I believe it will show you how it evolved into the major mine that it has become. I hope that you enjoy the story.

Chapter 1

BEGINNINGS

The land that we know as the Mayo district, in the central Yukon, began its formation more than 600 million years ago. It is the product of processes taking place on a grand scale over a time span that is hard to comprehend. The human presence here would be the last second on a twenty-four-hour clock representing the history of the world.

The land began its existence at the bottom of the ocean, but the area was finally thrust upwards. The North American continent moved west, forcing the Earth's crust beneath the ocean in front of it to slip beneath its inexorable progress. This unstoppable movement caused a pileup of the Earth's crust that has slowly formed the mountainous southwestern corner of the Yukon as we know it today. Before that, the mountains in the eastern part of the territory formed the leading edge of the continent.

This movement would not have been visible to the human eye, advancing mere millimetres per year. The sediments deposited in what is now referred to as the Selwyn Basin were formed over hundreds of millions of years. These deposits were transformed over that time, under intense heat and pressure, into quartz/mica coarse-grained metamorphic rock know as schist. Other forces played a major role in the sculpting of the landscape. Uplift raised sea bottoms to mountaintops and molten intrusions injected mineral-rich material into the overlying sediments, while lateral movement ripped the landscape along a seam that we call the Tintina Trench, which cuts diagonally across the Yukon and into Alaska. Over time, from 200 to 50 million years ago, the rock formations sheared and shifted until reaching their present state, where the deposits along the southwestern side of the split have moved to the northwest by about 450 kilometres in relation to those on the northeast side.

About ninety million years ago, the continental plate slid over the oceanic plate, forcing it down into the earth's interior under immense heat and pressure. These forces transformed the subducted rock to a state of molten granite, which was pushed upward through fractures in the earth's crust. Among other locations, this material can be found today in the Potato Hills above Dublin Gulch. According to the late Yukon geologist Charlie Roots, "The intruding granites were giant heat engines, forcing water in cracks in the surrounding rock to circulate by heating it to near or above the boiling point. Hot water, because it is acidic, dissolved the minute quantities of gold, silver and other metals naturally occurring in most rocks. From these solutions, metals separated out where chemical conditions changed, or the temperature dropped."[1] This process accounts for mineral deposits throughout the hills in the Mayo district, including a network of gold veins flanking the granite of the Potato Hills.

Another force that has shaped the landscape in more recent times is the glaciation that has occurred over the past 2.5 million years. The Reid glaciation, which lasted from 311,000 to 82,000 years ago, terminated in the vicinity of Dublin Gulch. The McConnell glaciation that followed, between thirty thousand and fourteen thousand years ago, did not cover Dublin Gulch, thus sparing the region from the scouring effect of the ice and saving the placer deposits in the valley bottoms. All the land west of the continental glacial advance was free of ice, and the lowered sea levels created a land bridge between North America and Siberia, opening up an unglaciated expanse that has come to be known as Beringia. It is during the latter period of Beringia that people migrated from Siberia and eventually dispersed into North America.

Today it is hard to imagine travel on the continent without roads and airports, especially in the north. The main routes of travel by foot over land and along water corridors connected areas where various resources could be exploited, while others connected peoples in adjacent areas for trading and social purposes. The Northern Tutchone–speaking people who lived in the Mayo area for millennia before Europeans arrived were connected to others of the Mackenzie River, the Gwich'in to the north, the Hän to the northwest and Southern Tutchone to the south. Today the people of the Mayo area are known as the Na-Cho Nyäk Dun (Big River People).[2]

It is not known when the first people reached the Stewart and McQuesten river basins, even though the area was not covered during the McConnell ice advance. Archaeological evidence suggests that the region has been occupied for thousands of years. Before Europeans arrived, the Na-Cho Nyäk Dun were intimately connected with the land they lived in. They fished, hunted, trapped and foraged for food seasonally, moving from one area to another to exploit various resources as circumstances dictated. There were no permanent settlements, although there were significant gathering places where they would meet at certain times of the year. Yukon anthropologist Sheila Greer explains the transient lifestyle: "During earlier times, the Northern Tutchone people were not settled in one locale, but ranged widely. The summer season saw the people gather at fish camps along the Stewart River to harvest salmon, using fish traps. The fish were dried and stored for later use. In the fall, the families spread out, to hunt and snare game. The meat was dried and stored for the winter season, at which time some families would gather at dependable lakes, where fish could be caught with spears or nets."[3]

Trade was important, and goods were exchanged between neighbouring groups in all directions. The first interaction with Europeans probably came indirectly when furs were traded for European goods through neighbouring First Nations, then later, directly with European traders. Representatives of the Hudson's Bay Company (HBC) established Lapierre House to the north as its first trading outpost in the Yukon around 1843–44, but several years passed before there was any direct contact with the people of the Mayo region. In 1848, HBC trader Robert Campbell established a post at the confluence of the Pelly and Yukon Rivers, but that lasted for only four years before being destroyed by the Chilkat Tlingit from the Pacific coast.

More direct contact with Europeans did not occur until the 1870s, when Jack McQuesten established a trading post at Fort Reliance on the Yukon River, a few kilometres downriver from what would later become Dawson City. In 1886, McQuesten and his partner, Arthur Harper, established a trading post at the mouth of the Stewart River in response to the discovery of placer gold in the Stewart River valley. From that time on, there were regular incursions of European prospectors into the region in search of placer gold in the bars of the Stewart River and its tributaries.

Prospectors and traders came into the Yukon during the last two decades of the nineteenth century before the first government officials arrived. They probed the Yukon River and its various tributaries, searching for placer gold in the tradition of gold-seekers in California, the Cariboo and the Stikine areas to the south. At first, they used the most primitive of methods to search for and recover gold. They lived in deprived circumstances, cut off from the outside world by the harsh weather that descended upon the region every winter. Starvation, scurvy and consumption loomed over them like an ominous dark shadow. Their improvised hovels lacked the luxury of stoves to keep them warm, or windows to give them light during the short daylight hours that came with the brutal, frigid winters. Their lives improved after supply lines were established for the goods they needed to continue their search for gold, and they spread out along the main branches of the Yukon River, and then to the tributaries of the branches.[4]

Gradually, the findings improved and more prospectors trickled into the Yukon year after year. Each of these men came with the dream that he would strike it rich. After gold was found along the Stewart River, better diggings attracted them to the Fortymile, then Sixtymile, and then the Circle district in Alaska. In August of 1896, gold was found in the Klondike valley in such quantity that it sparked a major gold rush into the region and led to the formation of a separate administrative territory now known as the Yukon in 1898.

The real impact of European colonization on the region's First Nations did not really begin until tens of thousands of people stampeded into the Yukon as the gold rush took hold. The European incursion had an increasing impact upon Indigenous peoples of the Stewart River through trade, a wage economy and the suppression of traditional cultures. Through legislation enacted by the Canadian government, the Department of Indian Affairs embarked upon a program intended to "civilize" the Indigenous people through education in mission schools, enfranchisement and introduction of British-style governance.

The first Indian agent arrived in 1906, and while Indigenous people contributed to the economy through trapping and hunting, these traditional activities were affected by market forces that dictated the up-and-down demand—and thus price—for furs. Construction of roads eventually eliminated employment on river boats and wood camps along the Yukon. Game laws were imposed upon Indigenous peoples throughout the Yukon, restricting their access to and use of their traditional food sources. Control of the land and its use was thereby taken away from them by the Canadian government, compounding other factors to inflict devastating results upon the people who had lived in the region for so many generations.

Meanwhile, it was the dream of mineral riches that inspired the newcomers, and mining spurred the development of the territory. It is upon this foundation that the Yukon economy thrives today, leading to the construction of the Yukon's largest gold mine: Eagle.

CHAPTER 2
PLACER

Placer mining in the Mayo district predates the Klondike gold rush. Prospectors had been exploring the Stewart River and its tributaries in earnest since 1885. In 1893, William Rupe and an unnamed partner went up the Stewart to the mouth of the Mayo River. The river, the lake and the townsite that later developed at the river's mouth were named after pioneer Al Mayo, who in 1885 operated a trading post at the mouth of the Stewart River. The prospectors ascended the Mayo River to Mayo Lake by poling boat, a common practice before reliable and regular steamer travel. The only place where they found any gold was on the bank of the Mayo River above the mouth of the present-day Mayo Creek.[5]

Three years later, in 1896, Thomas Nelson found gold four miles up a small tributary of the McQuesten River that he named Nelson Creek. Later that summer, Thomas Haggart built two cabins on Nelson Creek and another on a small tributary that he named Dublin Gulch. The two men did not stake the ground at that time. The following year, they were drawn into the discovery of gold in the Klondike district. The massive stampede into the Yukon Valley in the fall of 1897 and spring and summer of 1898 included hundreds, even thousands, of prospectors fanning out along the tributaries of the Yukon River, looking for more gold. Fearing that their ground on the remote tributary of the McQuesten River might be staked by newcomers, Thomas Nelson, Tom and Peter Haggart, and Warren Hiatt headed back to their remote gold prospect. A dispute developed en route to Nelson Creek, and the four men split into two parties. Peter Haggart and Hiatt reached the creek first, with Haggart staking the Discovery claim on June 23, 1898, and renaming the creek after himself. More prospectors subsequently arrived.[6]

One of the names that appeared prominently in early Dublin Gulch placer mining was that of Kentuckian John Jackson "Jack" Suttles. A veteran of the Spanish-American War, he arrived on Dublin Gulch in 1899, and in a period of eight days during his first season, he shovelled gravel into a sluice box and recovered $364 in gold dust.[7] Suttles was a big, ruddy-faced man and a musician of note, having performed with a minstrel group in the American South before coming to the Yukon. Whenever he arrived in Dawson City for business, or to pick up supplies, "Happy Jack, the minstrel-miner of Dublin Gulch" would mount the stage in the local theatres and perform. "Jack has an assortment of songs, and a flock of banjos and guitars," reported the *Dawson Daily News* in 1910, "... the everlasting smile and warbling trombone voice—all of which contrive to make him the best single-handed entertainer in the Yukon."[8]

Dublin Gulch was quickly staked for placer mining from top to bottom, as was Haggart Creek, the stream into which the gulch fed. On Haggart Creek claim No. 25, known as the "lost rocker" claim, a hole was sunk eight feet that was recovering twenty-five cents of gold to the pan, and they hadn't even reached bedrock.[9] There was a flurry of mining in 1899; Jacob "Jake" Davidson and others were sinking shafts on the benches and mining both Haggart and Dublin. This prospect work petered out within a couple of years, however, and many of the claims were abandoned.[10]

Haggart and Dublin were only the first gold discoveries in the Mayo district. Prospectors fanned out across the country and it wasn't long before discoveries were made on other creeks in the district. In 1898 three Swedes, a father and two sons named Gustafson, were prospecting up the McQuesten River and found enough coarse gold to pay for supplies in Dawson. For three years they had sunk shafts and drift-mined. They built a cabin and a sawmill powered by an overshot water wheel. All of their work was by hand; without the use of nails they used wooden pegs, and they "sheathed the bearings and axle of the water wheel with flattened coal-oil tins. They were very secretive about the location and refused to stake the ground."[11]

Jack Suttles's placer mining operation on Dublin Gulch, 1912.

That was the downfall of the secretive Swedes. In the autumn of 1901, a hunting party consisting of Duncan Patterson, Colin Hamilton, Jake Davidson and Alan McIntosh, who were mindful of what the Swedes were doing, encountered their diggings on what became Duncan Creek. Patterson and Hamilton staked a Discovery claim on September 15, 1901, and named the creek. Davidson staked No. 1 above while McIntosh took No. 2 above. The Swedes refused to stake here and moved on. That first winter, McIntosh rocked out sixteen hundred dollars in gold; one pan was said to have produced fifteen dollars. The following year, Patterson, McIntosh, Hamilton and Davidson took out about forty-five thousand dollars in coarse gold. More prospectors came into the area in 1903, but gold production fell to thirty thousand dollars, and only half of that in 1904.

Duncan Creek was claimed from end to end in the stampede that followed the staking by the original four. Shortly thereafter, as many as two hundred men were prospecting in the region. Duncan Creek had become the next big gold find after the Klondike. Jack Turner, who had claim No. 3 above Discovery on Duncan Creek, mined continuously for the next nineteen years. Future gold commissioner George P. Mackenzie and his wife, Thora, spent two years mining at Discovery on Duncan Creek at one time.[12] A settlement was established on the Stewart River at Gordon Landing in 1902 that provided access to Duncan Creek via a two-mile portage to Janet Lake and a twelve-mile stomp from there to the diggings.

Gordon Landing was not easy to reach by riverboat, however, and its population dwindled when the government surveyed a townsite near the mouth of the Mayo River.[13] Mayo Landing had the advantage of being a good location to land supplies and haul them overland to Duncan Creek. The townsite of Mayo Landing was surveyed in 1903 by Raoul Rinfret, Joseph Edward Belliveau and John Dease Bell. Hearing rumours of the future townsite, Alex Nicol built the first cabin there in March 1903. He was followed by Gene Binet, who began to build a hotel. The hotel was completed by July, before the surveyors arrived on the scene. Cabins quickly popped up at the site.

The North-West Mounted Police (NWMP) established a detachment in 1904, the same year that the government began the construction of an overland trail from Mayo Landing to Duncan Creek. An overland winter road from Dawson City that intersected the Duncan Creek road at Minto Bridge was completed in 1905.[14] Gradually, over the years, other businesses and services gravitated to Mayo Landing, and it became the commercial hub for the district.

While a government road to Duncan Creek reduced transportation costs significantly, such was not the case with Haggart Creek and Dublin Gulch. "Getting to Haggart

Creek in the summer involves a hard tramp of about fifty miles cross country from Mayo Landing," reported the *Yukon Morning World* in 1908. "The trail most generally used is that leading from the government road about a half mile below Fields Creek extending up Black Creek, thence across the divide and down Ross Creek to the wood jam at the mouth of Haggart from which point the trail is blazed to Dublin gulch." The first ten miles out of Mayo to the Minto Bridge were described as a good wagon road, and the next ten as rough, though passable, for a heavy wagon. After that, the trail was "impassable for any conveyance other than pack animals. Muskeg, marshy and hummocky ground are encountered."[15]

The logjam located at the mouth of Haggart Creek was noteworthy for its proportions. According to Hugh Bostock, a geologist with the Geological Survey of Canada who visited the area thirty years later, it was still in place and "completely filled the course of the river for at least a quarter of a mile, extending nearly down to the mouth of Haggart Creek."[16] Beyond the logjam were permafrost bogs that were only passable in the winter when supplies and equipment could be hauled in over the frozen ground. Despite the transportation difficulties, the mining on Haggart and Dublin persisted. Many of the claims staked in 1900 lapsed within a year and were restaked two years later.

Charles Rawlins, who mined on Dublin Gulch in 1902 and 1903, estimated that the gravels yielded three dollars per cubic yard of gravel sluiced.[17] In 1904, government geologist Joseph Keele reported, "Work has been carried out here every year since 1898, but only two men were working here during the past summer. They were engaged on claim No. 15 above Discovery. The work consisted of washing out the gravels in the valley bottom by means of a small hydraulic plant."[18]

Many of the original claims staked in 1898 were abandoned and became open ground again; then, in 1907, William and Nathan Abbott found gold along the left limit of their Haggart Creek claim in values as high as $3.75 to the pan.[19] Their winter dump was yielding forty cents to the bucket with little effort. Combined with the claims being actively worked on Dublin Gulch, the news was enough to spark another stampede to the creek. The following year, there were fifteen men on the creek; by 1909, there were more than fifty men working away on claims on Haggart Creek and Dublin Gulch.[20]

There was barely enough water on Dublin Gulch, reported the geologist Joseph Keele, for hydraulic mining.[21] Despite that shortcoming, James Haddock obtained a hydraulic mining lease for Dublin Gulch on August 30, 1905, excluding any claims that were still in good standing. The creek was not suitable for mining by any other means, he claimed, but

George Black, a lawyer and member of the territorial council, complained that such concessions tied up ground and prevented hard-working and enterprising individual miners from prospecting and staking wherever these concessions were granted.[22] The council, which contained a majority of government-appointed (non-elected) members, voted down Black's resolution of censure.[23]

The concession remained in effect and by December 1908 it had come into the hands of Dr. W.E. Thompson, a Dawson City physician, who hired hydraulic mining expert V.V. Blodgett to oversee the work of hydraulic mining. The Dublin Hydraulics, Ltd., was incorporated under the federal Companies Act on March 10, 1910, with Dr. Thompson serving as president and A.W.H. Smith, the secretary-treasurer, as the promoter and mover of the project. Dr. Thompson transferred the concession to the Dublin Hydraulics

A promotional image for Dublin Gulch in the *Dawson Daily News*, December 1910.

Hydraulicking and Prospecting on Properties of The Dublin Hydraulics, Limited

Hydraulicking Scene on Dublin Gulch   A Clean-up on Dublin Gulch   Ground Sluicing on Dublin Gulch   Prospecting Scene on Dublin Gulch

Company, Ltd. The concession was the sole asset upon which fifty thousand shares in the company were issued. Thirty-five thousand of those shares went to Dr. Thompson, who returned nine thousand of them to raise money to conduct the business of the company.[24]

A full-page spread promoting the new company was inserted into a special Christmas edition of the *Dawson Daily News* in December 1910. In it were photos depicting activity on the creek; preferred shares were on offer for the sum of $2.50 each by contacting Smith at the company office of T.A. Firth, a Dawson mining broker and fiscal agent. Steel pipe, lumber, provisions, horse feed and other supplies were shipped to Mayo Landing on the last boat before freeze-up of the Stewart River in October, and were being readied to haul to Dublin Gulch before the spring thaw. A ditch was to be constructed to divert water from Haggart Creek to Dublin Gulch in order to wash the gold from the gravels.

Jack Suttles, who owned ten claims within the confines of the concession, optioned them to the company for ten thousand dollars, to be payable in two years. Mining equipment was to be purchased and shipped to Mayo in 1911, then hauled on-site over the winter. Mining would commence in 1912. It was estimated that at seventy-five cents' worth of gold per cubic yard, the ground to be mined would yield $4,622,750 in gold. The profit would be $880 per day. Over a season of 150 days, the profit to investors would be $132,000 each year.

But the mining company had counted their nuggets before they hatched, and little more was heard of the Dublin Hydraulics Company.[25] No actual mining took place on the concession for several years. The concession was cancelled and opened again to staking on February 5, 1917. Despite temperatures between minus fifty and minus sixty degrees Fahrenheit, prospectors immediately staked all the ground except for four claims at the upper end of the concession, which were considered to be of no value for placer work.[26]

Mining continued at Dublin Gulch until 1914, when two circumstances changed the course of development of the creek. The first was litigation that took place between Jack Suttles and the Cantin Brothers—Francois ("Frank") Cantin, and his cousins Louis and Phileas—formerly from Trois-Rivières, Quebec. Suttles filed a suit for two thousand dollars against the Cantin Brothers, alleging that they "opened a flume gate which allowed water to fill in a cut, thereby interfering with the success of his operation."[27]

The Cantin Brothers counter-sued, alleging that Suttles allowed the tailings from his mining to flow down onto their ground, disrupting their operation. The cases went to court in October 1914, and the judge eventually found in favour of the Cantin Brothers. Sheriff Brimston seized Suttles's claims on Dublin Gulch, and they were sold. Eventually, they were purchased by the Cantin Brothers.[28] During the years that Suttles mined on

Frank Cantin (second from left) with his cousins Joseph, Louis and Phileas (in unknown order) on Dublin Gulch.
Frank, Louis and Phileas operated as Cantin Brothers on Dublin Gulch and Haggart Creek from 1909 to the 1920s.

Dublin Gulch, it is said that he recovered more than forty-five to fifty thousand dollars in gold from his claims.[29] Today, that gold would have been worth more than four million dollars. It is also said that Jack Suttles put a curse on the ground so the Cantins would never make a go of it; in reality, the ground was just too difficult for them to mine, due to deep overburden, large boulders that interfered with easy mining, inaccessibility and a chronic shortage of water.[30]

The other event that hampered the continued development of placers in the Mayo area was the outbreak of the "Great War," later to be renamed World War I. With the announcement that Britain had declared war against Germany (and shortly thereafter, Austria) in August of 1914, a steady stream of healthy, robust Yukon prospectors and miners stepped forward to volunteer in the Canadian Expeditionary Force. One of the first to do so was James Murdoch "Jim" Christie, in September 1914. Christie had frequented the Mayo district for several years, and a creek near Dublin Gulch is named after him. Suttles followed suit, enlisting in the autumn of 1916 and joining the Yukon Infantry Company that was assembled by George Black, the commissioner of the Yukon.[31] After the war, Suttles missed the opportunity to use his travel voucher when he was demobilized from the army in 1919, and never returned to the Yukon. George Crisfield, who was Christie's mining and trapping partner in the Stewart River area, enlisted in 1918 and was killed during the September 1918 assault of the Canal du Nord in France.

One of the deterrents to men joining up was the requirement for continued assessment work to keep the claims in good standing. Dr. Alfred Thompson, the Yukon member of parliament, lobbied for—and got—an exemption from this requirement for any men volunteering for military service for the duration of the conflict. Dr. Thompson went further to stimulate mining on Dublin Gulch. He lobbied the Imperial Munitions Board to establish an office in Dawson City to purchase tungsten ore. "The tungsten ore scheelite, found on Dublin Gulch," he said, "is worth about one thousand dollars a ton in Liverpool. The cost of shipping it from Ottawa to Liverpool is $232 per ton." Dr. Thompson was hopeful it would help stimulate the Yukon mining industry, but nothing came of it.[32]

Despite the wartime depopulation of the region, fourteen men were reported to be still engaged in mining on Dublin Gulch. Access to the creek still created problems for the miners, and although ten thousand dollars had been included in the territorial budgets in 1914 and 1915 for a road to Dublin Gulch, it wasn't built, presumably because of wartime priorities. During 1918 and 1919, a double-ender sleigh road was to have been built to the creek, but the war, rising costs, expensive transportation and a stagnant gold price all combined to diminish interest in mining on Dublin Gulch through this period.[33] Despite the improved access, the only miners who reported mining on the creek by 1925 were brothers Nathan and William Abbott, as well as William Portlock and Robert "Bobbie" Fisher. Curse or no curse, the Cantin Brothers never had luck with their Dublin Gulch claims and let them lapse in the late 1920s.

Keno City grew up in proximity to several hardrock mines that had been developed in the vicinity of Galena and Keno hills, fifteen miles to the southeast of Dublin Gulch. Yukon Gold Company built a mess house and stable at Keno City, as did other mining companies. Harry Yamasaki was put in charge of the roadhouse there, eventually becoming the proprietor. By 1922, Keno City had a post office, a government assay office and a Mounted Police post. There were at least four hotels and several general stores, including one operated by returned war veteran Norton Townsend. Thomas Jackson and Arthur Major operated a poolroom and barbershop. Miss Jessie Stewart ran a novelty store. At least thirty cabins were built at Keno City, with more down in the valley or on nearby hill claims. The unsurveyed townsite was laid in disarray, as if a child had thrown down a handful of giant Lego blocks on the landscape.[34]

By 1929, the town included the Northern Commercial Company store managed by Dick O'Loane, the mercantile business Taylor and Drury (known universally as T&D), a small bakery run by Mrs. Erickson, Jimmy Sugiyama's hotel and café and Jackson's Saloon, with a hall on the top floor that operated as a social club. There was a post office (with "Tiny" Greaves as postmistress), the Royal Canadian Mounted Police (RCMP) under Corporal Coleman, the liquor store run by Vic Grant and the government assay office (with Billy Sime as assayer and government agent). In addition, there were a few cabins upstream on Lightning Creek which housed a number of women with no 'visible means of support.'

Theodore "Ted" Bleiler was a schoolteacher from Alberta who began teaching at the school in Keno City in the fall of 1929. It was by pure chance that he met a former classmate who told him of the higher rates of pay offered to teachers in the Yukon, compared with those in Alberta. One teacher, he was told, saved enough money from his Yukon wages to purchase a farm. Bleiler began to teach in the tiny northern settlement, taking the summers to work and prospect in the adjacent hills.

The summer of 1932, when school was out, Bleiler and Norval Lochore did some assessment work for Bobbie Fisher and Archie Martin on tungsten claims they held in

Ted Bleiler on his claim during the 1930s.

the shadow of the Potato Hills overlooking Dublin Gulch. Bleiler and Lochore visited the Olive hardrock mine and continued down to the placer diggings abandoned by the Cantin Brothers, where they were told they could pan for gold. They took up temporary residence in an old cabin beside the open ground. Their nearest neighbour was a miner named Jim Gibson, who had settled in for a "drying out" spell.

For three weeks Bleiler and Lochore sluiced the ground, taking out fourteen hundred dollars in gold. When they weighed out the gold they had recovered, there were several spectators, including Ed Barker, Kit Watters, Jack Hawthorne, Louis Kazinsky, Jim Gibson and Archie Martin. Bleiler and Lochore staked a two-mile prospecting lease. Their cleanup was enough to start another staking spree, with Jim Gibson getting the claims on Haggart Creek at the mouth of Dublin Gulch. Bleiler bought the mining equipment left by the Cantin Brothers for $150.

Bleiler gave up teaching that year and turned to mining. During the winter he, along with Bobbie Fisher and Norval Lochore, mined their property, but the rich ground that got them started petered out. Fisher gave up after the summer of 1933 and Lochore left in 1936 to take up fruit farming in the Okanagan region of British Columbia.[35] Ted Bleiler ended up as sole owner of the Dublin Gulch claims; he worked the property by himself for one year, then took on a partner named Fred W. Taylor in 1937. Bleiler married that year, and after one last shot at struggling with water shortages, poor ground and isolated conditions, he abandoned his dream of golden riches and sold the ground to Taylor.

Fred Taylor was a beneficiary of the Great Depression of the 1930s. He left school after completing grade ten, but jobs were hard to come by. He heard that there were jobs mining silver at Elsa and Keno City, so in the spring of 1936 he caught a freighter to Skagway, took the train from Skagway to Carcross, then walked all the way to Mayo to save his last few dollars. He did not take his pack off from Carcross until he reached Montague roadhouse, 246 kilometres later. During this stretch he occasionally rested by leaning against a tree, but never took a "real rest." Since it was springtime, he had to cross three major rivers as they were "breaking up" by leaping from one ice chunk to another. (For precaution, he had a long pole he carried and could use to pry himself out of the water on ice chunks if he fell into the river.) This determination gave him a head start on the men waiting for the ice to go out so that they could cross on a raft.

After almost two years of not finding work in the south, Taylor was determined to find employment in the Mayo area.[36] He couldn't find steady work when he got there, so he supplemented his diet by trapping rabbits and squirrels. He secured a job hauling provisions up the Stewart River to Lansing Post for some trappers and hauled the provisions of another trapper back to Mayo in exchange for half the load.

Taylor eventually got work loading sacks of ore concentrate onto the river barge that was pushed by the steamer *Keno*. They handled these heavy loads onto the barge by two-wheel dollies. The dollies each had four to six gunny sacks on them, and each sack weighed 110 to 140 pounds.[37] At the end of the season, everybody without full-time employment left for the winter, but Taylor stayed because he loved the country. Within a couple of months, the manager of the mine at Keno needed someone with a steam ticket to run some steam equipment. Taylor tapped into his previous experience in hardrock mining at Bralorne gold mine, west of Lillooet in British Columbia, and kept the boiler running all winter.[38]

In the spring of 1937, Taylor heard of Ted Bleiler and his placer mine on Dublin Gulch. He hiked to Ed Barker's claim on Haggart Creek, where he was offered work, but he pressed on to Dublin Gulch. He spent the rest of the summer working for Bleiler. Together, they worked four thousand cubic yards of ground and recovered about one hundred ounces of gold and several hundred pounds of scheelite. Bleiler subsequently turned the Dublin Gulch property over to Taylor, who continued to mine there for several decades.[39]

Taylor was at Dublin Gulch all by himself on the claim that winter, bringing in a supply of wood. He was forty-six miles from Mayo with no road—only a foot trail—no radio and no help. He was resourceful, but also careful. He decided to sink a shaft where, by studying the terrain, he thought he would find gold, and reached bedrock about twenty-five feet down. He broke up the large rocks with a hammer, sometimes using fire to crack the boulders and reduce them to more manageable size, and drifted two directions from bedrock to determine the better gold location.

The summer of 1938, George Potter and associates hauled a considerable amount of equipment into Lynx Creek, a tributary of Haggart just below Dublin Gulch, where several hundred feet of ditch were repaired and a large flume and sluice box were constructed, but the effort was closed down in the middle of the summer. On Haggart Creek, well-known prospectors Tom McKay and Archie Martin were working the gravels.[40] Meanwhile, Taylor and two other men worked a cut where there was ten to fifteen feet of gravel over bedrock. Prior to Taylor owning these claims, no one had placer mined to bedrock. They had been mining surface gravels only, which yielded a grubstake for the winter.[41]

Taylor started mining on bedrock, but government geologist Hugh Bostock reported that the gold was scattered throughout the gravel, not concentrated at bedrock. Bostock reported that it consisted of "a layer of compact, angular fragments of local bedrock cemented with a hard, sticky clay on bedrock and seems to act like a false bedrock."[42] In addition to gold, there were other heavy metals at Dublin Gulch like scheelite, a tungsten ore that tended to clog the sluice box and had little economic value at the time. There was very little water during the summer, so Taylor built a retaining dam. He directed the water using flumes to control the flow of the water against the mine face. The sluice box would be buried in the bedrock. The coarse material washed through the apparatus, but the sluice captured the gold.

To move the heavy boulders out of the way, he built a derrick to hoist them up. The derrick was held in place by four cables anchored to deadmen. He saved up enough money to purchase a hand-operated one-ton hand winch to help him. He would move the derrick ahead by winching it slowly forward. The bottom of the mast was placed in a bucket of grease. The mast was tilted slightly so that once the boulder was winched up, it would swing to the side naturally. His goal was to move one hundred loads of rock a day, about one hundred tons, seven days a week all summer long. Taylor hired two men to help him; between the three of them, each man moved thirty-three tons of rock every day. All this hard work was done with a limited amount of mechanization, an amazing feat at that time.[43]

In 1939, Fred Taylor continued to mine Dublin Gulch, now with three men and a gasoline-powered winch added to his derrick to improve the operation. They processed ten thousand cubic yards of gravel and recovered 325 ounces of gold, along with another heavy metal, tungsten, from his sluice box.[44] The tungsten was found mostly in the form of scheelite; it was estimated by Hugh Bostock that ten tons of scheelite concentrate could be produced per year if there was sufficient water for the work to be done.[45]

Taylor went to Vancouver in the winter of 1939–40, where he met his future wife, Ann, a hairdresser. She was familiar with the wilderness lifestyle; her father had been a

Fred Taylor's rock derrick on Dublin Gulch. The derrick served to remove the rocks and boulders that interfered with the mining.

Fred Taylor's cabin
on his placer
operation on
Dublin Gulch.

trapper and trader in northern BC in the Tintina Trench area. They married in the spring
of 1940 and returned to Mayo. Fred came back to Mayo first, and built a house by hand at
Dublin Gulch in three weeks. It became their family place for many years and its remains
still survive today.

They began to live at the creek year-round. According to Fred's son Frank, these were
happy times. Fred skied in the winter; Ann took care of the house. Dublin was substan-
tially warmer than Mayo because of the elevation, and there was not much wind. The
house was comfortable, requiring seven cords of wood per year to heat it, mostly cut by
hand on a swede saw. Fred converted an old cabin into a sauna, which they would enjoy
in the winter. They heated rocks and poured snow on them to produce steam.[46]

Two events occurred during the summer of 1939 that had an impact on mining on
Dublin Gulch, though one had greater consequence than the other. First, long-time
prospector Bobbie Fisher died after having spent several decades mining in the vicinity.[47] Then in September, Canada joined Britain and other Commonwealth countries in
declaring war with Germany. The conflict would last for six years.

Suddenly, the scheelite that had been a nuisance clogging sluice boxes became a mineral of strategic importance. Because of its density, it could be used in place of lead, but more importantly, its hardness and high melting point are properties that make good armour-piercing projectiles, as well as protective armour plates. But extracting it and getting it to where it would be of value to the war industry was hampered by poor transportation.[48] The route from Dublin Gulch to Mayo consisted of a winter road from the gulch to the logjam on the McQuesten River. That road would have to be improved, and bridges constructed to span the McQuesten River and Haggart Creek. But during the summer of 1940, that was still a speculative dream.

An example of the difficulties of transportation that plagued Haggart and Dublin was the accident in 1940 involving Peter Gatey, who was working for Taylor. Gatey was seriously injured when his pick bounced back and hit him. Another of Taylor's employees, Jack Shandro, administered first aid and then a sled hauled by Ed Barker's Caterpillar tractor could take them only as far as the logjam on the McQuesten River. From there Taylor, Shandro and five other men carried Gatey by foot on a stretcher. With the stretcher mounted on their shoulders, they stumbled over tussocks in the swampy ground. Peter's brother John went ahead of the men to the roadhouse located halfway between Mayo and Keno City and telephoned Mayo for help. Dr. Geoffrey Homer came out from Mayo to meet the party two miles beyond the roadhouse, and Gatey was transported from there by truck back to Mayo where he underwent emergency surgery.[49]

Inspired by the growing success of the mining by Taylor and Ed Barker, a number of other men moved into the area to try their luck, including Irvin Ray, Vilhelm "Ole" Lunde, Robert Swanson, Clifford Greig and Hugo Seaholm, all of whom were testing the gravels of Dublin Gulch.[50] Hugh Bostock made his annual visit to the creek in 1941 and talked to Taylor and Barker about the prospects of mining more scheelite. Taylor was operating his mine with a crew of four to five men, while Barker was operating a well-financed concern on Haggart Creek, equipped with a bulldozer, dragline scraper, front-end loader, steel sluice boxes, welder and other repair equipment. This was the first highly mechanized mining operation to work on Haggart Creek.

Included in Barker's crew was a cook, Kate "Kay" Broadfoot, who was not only good at her job, but charismatic and educated as well. She was not the only woman on the Gulch: Ann Taylor, Alberta Seaholm, Elsie Ray and Mrs. Robert Swanson joined their husbands on their mining claims during the summers.[51]

Fred Taylor's placer operation on Dublin Gulch.

With the war in its third year, tungsten was becoming extremely important to the federal government. In the spring of 1942, the government financed the completion of a bridge over the McQuesten River and a road to Haggart and Dublin. Fred Taylor had a good mining season, extracting seven hundred ounces of gold and four thousand dollars' worth of tungsten concentrates. Seaholm, Lunde and Swanson produced more than two tons of scheelite concentrate. The new road to Dublin Gulch made delivery of the ore less expensive. Production of scheelite increased dramatically in 1943 when the price went up by 50 percent, and the government began paying 75 percent of the value of the concentrate once it reached Mayo. [52]

Combined with other mining going on in the region at the time, Mayo was thriving. The three mercantile operations in Mayo reported that business was much better than expected, in part because of the wartime economy and especially because of the friendly invasion of Americans. After the bombing of Pearl Harbor in December of 1941, the United States entered the war. The Japanese, meanwhile, gained a foothold in North America when they invaded the Aleutian Islands off the coast of Alaska. This threatened the American shipment of strategic material along the Alaskan coast, so the United States undertook to construct an all-weather road that would connect Alaska with the continental United States. A second massive project was the building of a pipeline from Norman Wells in the Northwest Territories to a refinery in Whitehorse that was to produce much-needed fuel for the military in Alaska. The need for labour and supplies increased dramatically.

Fred Taylor on leave in 1943 with his wife, Ann, and new son, Frank. Fred later served in several theatres of battle in Europe.

The two sawmills in Mayo sold their complete output to the United States Army and shipped it to Whitehorse.[53] Bostock described Mayo at this time as congenial: "The men were making money and the ladies, besides their teas, had berry-picking picnics and evening badminton parties on the court Irvin Ray made."[54] George Black, MP for the Yukon, made a point of visiting Mayo and reported that social and church activities were continuing as usual, "while the volume of bundles of knitting and sewing sent out by the IODE [Imperial Order, Daughters of the Empire] is positively surprising."[55]

Fred and Ann Taylor went outside (left the Yukon) in the fall of 1942 and Fred joined the army.[56] Before Fred was shipped overseas, his first son, Frank, was born in 1943. Over the four years Fred was away, he leased his property to two other men, one of them being Ole Lunde. Fred served in the Canadian tank corps and was reportedly stationed in North Africa, Sicily and Italy, and then France, Belgium and Holland after the Normandy invasion. Once he was demobilized, Fred and Ann quickly returned to Mayo. Two other men who enlisted from Dublin Gulch never came back from the war.

In 1946 or 1947, Taylor spent the money he received from leasing his claims on a brand-new D7 bulldozer, the first one in the area to use hydraulics to lift the blade.[57] His son Frank remembers his father was always very careful with his money, never spending it on the wild frontier lifestyle so typical in remote mining towns.[58] The bulldozer was a real boon to mining compared to pick and shovel; it was good for cleaning bedrock, digging ditches and clearing ground. In 1946, employing a crew of five, Taylor recovered 608 ounces of gold. In 1948, using a small crew and power equipment, he recovered 500 ounces.[59]

In the winter of 1949, the Taylors moved into Mayo so that Frank could attend school.[60] With a stagnant gold price and rising expenses, there were better ways to make a living. There was very little mining in the district and none on Dublin Gulch until Taylor returned in 1953. In 1951, however, during the Korean War, he was able to ship out scheelite from a stockpile that had accumulated on his claim from previous years of mining, and enjoyed a return of fifteen thousand dollars for his efforts.[61] When he began mining Dublin Gulch with a bulldozer again in 1953, he recovered 549 ounces, as well as concentrates of scheelite, which he accumulated until he had enough to ship out at a profit. Fred Taylor recovered 450 ounces in 1954, and then sold half his Dublin Gulch claims to Clifford Greig in 1955.[62] Unfortunately, a boulder crushed Greig's leg in 1959, so he sold the property on to George Smashnuk, who continued to mine the ground until 1962.[63]

Also in 1955, Ed Barker leased his Haggart Creek property to Waddco Placers, Ltd., which was a partnership of Maynard Wilson, Jack Acheson, Bill Drury and Clyde Day. They employed a crew of five or six each season until 1959.[64] Using two D8 Caterpillars and other heavy equipment, they mined on a scale that dwarfed Fred Taylor's efforts on Dublin Gulch. In 1955, they recovered nearly 4,500 ounces of gold, almost ten times the amount Taylor recovered on the adjacent claims the same year. By 1959, when Waddco moved to placer ground on Dominion Creek near Dawson City, they had reportedly recovered $400,000 in gold from the Haggart Creek property.[65]

Installing a sluice box at Haggart Creek.

Left: Clean-up, Haggart Creek.

Right: Operating a cable winch mounted on a Caterpillar tractor, Haggart Creek, 1950s.

Taken during placer mining at Haggart Creek in the 1950s, this photo shows a Caterpillar tractor pushing dirt toward a dragline bucket.

Moving pay dirt
by Caterpillar,
Haggart Creek.

Dumping pay dirt into a sluice box, Haggart Creek.

Ed Barker's placer operation. A front-end loader is feeding paydirt into the sluice box, while a dragline (foreground) is removing the tailings from below it.

Ed Barker's placer operation below Dublin Gulch on Haggart Creek.

Placer crew in
front of cabin,
Haggart Creek.

In 1959, Ed Barker mined for a short while on Haggart Creek, recovering 123 ounces, while Fred Taylor's production on Dublin Gulch nearly doubled that year to 887 ounces. Barker leased some of his claims to Walter Malicky. According to geologist Aaro Aho, Malicky's prodigious efforts "exceeded his ability to coordinate the operation and make ends meet." Malicky was mining in an inefficient fashion, with cleanups every couple of days so that he could keep ahead of his creditors. Barker eventually took back his property.[66] Barker suffered a stroke at his Haggart Creek property in 1961 and died en route to hospital.[67]

Frank Taylor, Lowell Bleiler and Jim Taylor practising their skills as placer miners, Dublin Gulch, 1954.

Ed Barker had been a dredgemaster on one of the enormous mining machines in the Klondike goldfields when he was only twenty-three years old. For a period of time, he also worked on a dredge in Siberia. He did the same on Highet Creek, and worked for the Treadwell Yukon Company Ltd. at Wernecke Camp for eight years. He staked claims on Haggart Creek in 1932 and began mining there. By 1940, he held twenty-two placer claims and twelve miles of prospecting leases along Haggart Creek. He purchased the Chateau Mayo Hotel in Mayo in 1952 and later, the Tourist Services in Whitehorse, one of the Yukon's first supermarkets. After Barker's death, Jack Acheson acquired the prospecting lease above Barker's cabin on Haggart Creek in 1963 and mined it unsuccessfully for several years, selling the property before he went broke.

Howard White, who was a university student at the time, got a job working at Acheson's mine through Maynard Wilson from Waddco. White worked for seven months at the mine as a welding mechanic and catskinner. "I was the only one who stayed to close the camp," he recalls. "Doing so caused me to miss a year of university and it was sort of in a lost cause because the mine was broke and Jack was out of it with booze. But I never regretted doing it.

"In a rare moment of confidence, Jack told me I had saved the season and was 'a good man,' which was about the highest praise you could get from an old-time northerner, alcoholic and Nixon-lover that he was. I wore his grudging approval back south like a badge of honour and ever after thought that if nothing else in my life worked out, I could

The Taylor family on Dublin Gulch in the early 1950s.
Clockwise from top left: Frank Taylor, Lowell Bleiler, Jim Taylor, Fred Taylor and Marjorie, Fred's visiting sister.
Marjorie was a nurse who lived in Nelson, BC, for many years.

always go back north and be a placer miner. Actually, Jack offered me the mine in lieu of wages but I turned it down. Bad move—shortly after, they deregulated gold and it went from thirty-seven dollars an ounce to something like two hundred dollars." White never went back to placer mining, instead pursuing a successful career in book publishing. He eventually got paid for the summer he worked at the mine.[68] Acheson sold out to Klaus Djukastein, who mined the property in 1969 and 1970 before he, too, gave up.[69]

As Fred Taylor's sons, Frank and Jim, grew up, they joined him in his mining work. Frank remembers that he started working for his father at nine years of age, in 1952. He was paid eight and a half cents an hour. The half-cent was included, Frank believes, so that he could improve his math skills. When Frank was about thirteen years old, he started working machinery. A few years later, he spent a summer working for Jack Acheson on the Waddco claims on Haggart Creek. The crew worked ten-hour shifts, and there were two shifts a day. He remembers being paid $2.70 an hour just like the other men on the crew.

Fred insisted that his sons get a decent education and sent them outside to attend St. George's, a preparatory school in Vancouver, for their high school years. Frank followed up by going to university in Seattle, though he did not consider himself to be an exceptional student. But the call of the north was irresistible. When the price of gold started to rise in the 1970s, Frank followed in the family tradition. With financial support from his wife Bonnie's family, he moved his family to the Yukon so he could begin placer mining on Duncan Creek. His son and daughter helped when they were old enough, just like he did when he was young. His son, Troy, continues in the family tradition today.[70]

Fred Taylor mined his Dublin Gulch property for more than thirty years, but ceased after the 1970 season. The following year he sold the property to Ron Holway and Holway's good friend Darrell Duensing. Each of them invested twenty thousand dollars toward the down payment on the purchase of the claims, with final payment the following year. Their deal with Taylor was made with a simple handshake, which was the way things were done back then. Taylor, who had mined on the creek for decades, also kept his gold sorted according to nugget size in large jars in his cabin, and never worried about being robbed.[71] Soon after the sale, Taylor moved outside with his new wife, Joyce, a nurse from Mayo, and they started a family.

Holway and Duensing worked the property under the corporate name of Darron Placers Ltd., purchasing a D7 Caterpillar tractor that was in nearly new condition, as well as a Bucyrus-Erie bought from the City of Whitehorse. They also purchased a massive old cable-operated dragline as surplus from the Yukon Consolidated Gold Corporation. According to Holway, they had great difficulty loading and transporting the machine to Dublin Gulch. They eventually unloaded it from their truck and walked the machine the final leg of the journey. It was too wide for the tote road to Dublin Gulch, and only proceeded with great difficulty. It is rumoured that Duensing, an American, was worth millions of dollars; however, he enjoyed the hands-on work of operating heavy machinery.

Darron Placers worked the Dublin Gulch property for several years; it was very much a family operation. At first it was just the two partners working the claim, with Holway's wife, Helen, cooking and ordering parts. Helen would help at the mine on weekends, but would remain in Whitehorse during the week with their three children until school was out in June. Helen cooked food for them on the weekends, including filling up the cookie bin with chocolate chip cookies, which she said never lasted very long. Her procurement work frequently required trips to Whitehorse and back in one day for parts to keep the equipment operational.

Their sole employee was Ron Mullins, whom Holway considered to be an extraordinarily honest and reliable individual. They continued to find ways to improve their gold recovery. In 1977, working with a crew of three hired men, including Don Edzerza from Dease Lake, they improved recovery by using a subsidiary sluice at the lower end of their main sluice that received heavy fine gold not caught in the main riffles. An upward pulse or jigging action applied to the fabric bottom of this secondary sluice from a cam drive below allowed the efficient separation of gold from scheelite, of which six tons was recovered that season.

Simon Mervyn, who later became chief of the First Nation of Na-Cho Nyäk Dun, worked for Djukastein for several years, then for Holway and Duensing. "They made you work for your money in those days," he recalls. "You had to work. There was no welfare." Mervyn remembers when he worked on their Dublin Gulch property, they used Fred Taylor's old cabin as a bunkhouse: "I've never seen so many mice in my life as were in there. I had a twenty-two revolver, and you could buy those shot shells. I would get out of bed and a mouse would run across and... *boof*! We had four pails and all night they were falling in: a five-gallon pail with a little water in the bottom, and you put a rope across it with a piece of garden hose or something, bigger than the rope itself, and you leave it in the middle, right, put peanut butter on that, and they go across the rope and as soon as they get to that it spins, eh? *Boof*! You could hear that all night. *Splash*!"

Derocker and sluice plant used from 1984 to 1986 on Dublin Gulch. The gold channel mined during this period was very rich; the best cleanup was nine hundred troy ounces over four days of sluicing.

In 1977, a geologist named Gordon Gutrath had been invited by friends to start a new public company called Queenstake Resources Ltd. They were looking for minerals that were rapidly increasing in price. First, they staked a block of claims on tungsten mineralization on a tributary of Haggart Creek. They also optioned placer claims on Dublin Gulch. In 1971, gold, which had been set at thirty-five dollars an ounce, was allowed to float in value, and the price began to rise. By 1977, it had reached $161 and was attracting a lot of attention. "Our first exploration effort was in 1978 on the Dublin Gulch gold-tungsten placer property," relates Gutrath. "Our downstream neighbours were Ron Holway and Darrell Duensing, who were operating a successful placer mining operation. They were very helpful as we were real amateurs when it came to placer mining. One thing we did provide them with was lots of laughs."[72]

Ted Takacs with a pan full of gold at Gill Gulch.

With a crew of three, a D7 bulldozer, a front-end loader and a sluice plant that he had trucked in from British Columbia in the spring, Gutrath processed up to fifty yards per day, "depending on breakdowns." They shut down the operation in August 1977. That fall, the president of Canada Tungsten (Can Tung) met with Queenstake Resources and purchased a 30 percent interest in the company. They subsequently purchased the claims owned by Darron Placers Ltd. and then staked a large block of quartz claims covering Dublin Gulch. For the next few years, Holway and Duensing pursued placer mining opportunities on Bear Creek, another tributary of the McQuesten River, and then Reed Creek in the Kluane district in the southwest Yukon.

The summer of 1979, Canada Tungsten brought in and assembled a processing plant worth $1.5 million, which was capable of processing fifteen hundred yards of gravel per day. The results of a test run of the new plant in September 1979 looked promising. In 1980, they brought in more heavy equipment. Between June 15 and September, two crews working twelve-hour shifts processed 120,000 cubic yards of material. The following summer, a crew of twenty-six working two shifts began mining on a large scale using

Front-end loader feeding sluice on Ted Takacs's
placer claim on Gill Gulch..

three bulldozers, two front-end loaders, a backhoe, a scraper and two large Mack trucks. The results were disappointing, and they did not continue the work in 1982.

Instead, according to geologist Brian Lennan, who worked for Can Tung at the time, the company did extensive recirculating drilling where Ron Holway stopped mining. A zone of gold-bearing gravel was found near the mouth of Suttle Gulch. In 1983, with their tungsten mine closed, a small crew of three people that included Lennan stripped over-burden off the drilled area using three monitors and a D8 bulldozer. Mining commenced in 1984 and continued until 1986. Canada Tungsten recovered approximately ten thousand troy ounces of gold, which was ninety-two fine. Over those three years, Lennan poured thirty-three gold bars. The channel was completely mined out except for a small piece of ground.[73]

Since the beginning of placer mining on Dublin Gulch, 17,500 ounces of gold had already been recovered. After disappointing results in their attempt at placer mining on Dublin Gulch, Canada Tungsten sold the property back to Gutrath, who in turn leased claims back to Duensing and Holway, who began mining there again in 1988 under the corporate name of Dublin Gulch Mining Ltd.[74] Between 1988 and 1992, they worked with crews ranging from six to eight employees working two shifts daily.[75]

Duensing left the partnership after seven years to pursue other interests. To accommodate the crew, they brought in ATCO trailer units for a bunkhouse, and another unit for a mess hall. In 1993, Holway shifted his operation down to Haggart Creek and began mining the left limit below Platinum Pup, the next tributary feeding Haggart Creek below Dublin Gulch. They continued to mine the claim until 1999. The following year, the licence was decommissioned.[76]

Between 1978 and 2002, a few other placer operations continued to mine on Haggart Creek and its tributaries. Ted Takacs mined continuously on Gill Gulch until 2002, while Jack Frank and then later Rod Ramey mined on Fisher Gulch, opposite and above the mouth of Dublin Gulch. Victor Sharman ran a placer operation on 15 Pup during the late 1980s and until 1992. After the beginning of the twenty-first century, there was a resurgence of placer activity on Haggart Creek and its tributaries. In addition to those already named, Harry Johnson began mining on Murphy's Pup and Frank Plut on Swede Creek, while Keith Dye and Orest Curniski were testing ground on a two-mile prospecting lease higher up on Haggart Creek.[77]

But the placer mining of Dublin Gulch was complete. As of the closure of Ron Holway's operation in 2000, placer mining ceased on this tiny tributary. For more than

Ted Takacs panning the creek in his later years.

a hundred years, it had seen continuous activity, starting with Jack Suttles, followed by the Cantin Brothers, Ted Bleiler, Fred Taylor, Canada Tungsten, Gordon Gutrath and, finally, Ron Holway.

At first a remote and isolated tributary of a tributary of yet another tributary of the mighty Yukon, Dublin Gulch was far from the hub of placer mining around Dawson City. Yet, first worked by hand and followed by increasingly large-scale and sophisticated technologies, it continued to be a steady producer of placer gold. Wives joined their husbands and families grew up there. Today, there is a well-developed all-weather road allowing easy access to the gulch. Satellite links provide instant communication with the rest of the world. It is a far cry from the beginning of the twentieth century. While the placer mining sustained the tributaries of Haggart Creek for more than a hundred years, there was always the hope that someone would strike the motherlode on the rocky hills overlooking the valley. Quartz prospecting accompanied the placer activity in the shadow of Dublin Gulch, but it would be a hundred years before the hardrock potential would come to be fully realized.

CHAPTER 3

HARDROCK

Placer mining has traditionally been the "poor man's" mining method. While still expensive, it can easily be financed by individuals, families or small partnerships. Depending upon the concentration of loose mineral components within the deposit, extraction of minerals can be very rewarding for those engaged in this form of mining. Hardrock or quartz mining is a different matter. All stages of the work demand a greater investment of time and money in order to advance the development of the property, but the rewards, if successful, are enormous. Finding the motherlode is the stuff of dreams, the vision of wealth and success. From the earliest days of mineral exploration in the Yukon, prospectors have had visions of finding the next great mine.

The first quartz staking in the Dublin Gulch area was reported to have occurred in 1897, when Major S.C. McKim observed stakes for a quartz claim placed at the head of Johnston Creek, a tributary of the McQuesten River.[78] We know for certain that there was a staking rush to the area in 1901. On September 15 of that year, James Corkery staked the "North Star" quartz claim on Dublin Gulch, followed quickly by George Ortell the next day, then Thomas Haggart on September 19 and then Jake Davidson on October 4.[79] Interest quickly dwindled and many of these claims lapsed, but another flurry of staking occurred two years later.

By 1902, there were said to be two hundred men prospecting or mining in the Mayo district.[80] Corkery staked the "Barrett" quartz claim half a mile below the mouth of Dublin Gulch on April 28, 1903, and entered into a partnership with Harry McWhorter, J.A. Jake Davidson and W. Williamson almost immediately after that.[81] McWhorter would later play a significant role in the development of mining in the Mayo district. Among the others who staked in 1903 were J.K. Gordon, Thomas Edward Heney, John M. Davis, Hugh Corlin, William Morley Berry, G. Jones, Thomas Nelson Johnston, Cyril Vernon Henson and Jim Christie.

Most notable among these men, but not for his mining, was Christie, who became famous for an incident that took place six years later. Christie was born in Perthshire, Scotland, on October 22, 1867. Before he joined the stampede to the Klondike during the gold rush in 1898, he had been farming in Carman, Manitoba. He remained in the Yukon after the gold rush and in 1907 and 1908, guided Canadian geologist Joseph Keele on two expeditions into the region east of the Stewart River. On one of these trips, Christie successfully took Keele through the mountains to the Mackenzie River. The geologist spoke highly of Christie's abilities, and named a pass and one of the peaks in the Selwyn Mountains in his honour. A creek near Dublin Gulch was similarly named for Christie.

In October of 1909, Christie encountered one of the most blood-chilling challenges that faces anybody in the Yukon, when he was mauled by a grizzly bear. He and his partner, George Crisfield, were trapping on the Rogue River, a remote tributary of the Stewart River. Christie had been tracking a large grizzly bear that had disturbed one of their caches. The bear surprised him as he climbed up a snow-covered riverbank. At a range of thirty metres, he got off one shot from his Ross rifle, which hit the bear in the chest, and a second round to the head just before the bear was upon him. Christie tried to escape the charging grizzly, but to no avail.

The grizzly took Christie's head into his powerful jaws and began to crush his skull. Christie's jaw and cheek bone were fractured and his scalp was ripped away from his head, drenching the snow with his blood. One eye was blinded. To protect himself, Christie thrust his right arm into the angry bear's maw, and it too was crushed. Christie might not have survived had the bear continued its attack, but the bullets finally took effect and the beast rolled over, lifeless. Christie was in terrible shape. He was bleeding profusely, and his broken jaw hung open. He wrapped his jacket around his head to hold the fractured jawbone in place, and staggered half-blinded toward his cabin, which was eleven kilometres away. Leaving a trail of blood behind him, he made it back to the cabin.

Christie kept a fire going despite being half-delirious, until his partner, Crisfield, returned to the cabin. After a couple of days' rest, Crisfield strapped Christie into a sled and headed to Lansing, the nearest trading post on the Stewart River. Wrapped in his blood-soaked clothing, Christie endured in silence the pain from every bump and jolt on the four-day journey. For two months, trader Jim Ferrell and his wife, Helen, tended to Christie, slowly nursing him back to health. Ferrell even trimmed the jagged edges of Christie's scalp wounds as the flap of skin healed. Eventually, Christie was fit enough for the journey to Dawson. He, Ferrell and Crisfield left Lansing on New Year's Day. Christie even insisted on doing much of the physical work of preparing camp and leading the dog team on the journey to Mayo, and then to Dawson City, where he arrived in mid-January.

By this time, his jaw had healed improperly so he could not chew solid food and was reduced to consuming a liquid diet. The staff of St. Mary's Hospital in Dawson could do nothing for him, so he went to Victoria, BC, where surgeon C.M. Jones reset his jaw in the course of several operations. Dr. Jones told Christie, "You have no business to be alive." Much of the credit for his recovery goes to the Ferrells, who tended to him for so many weeks. From that time on, he was known as "Grizzly Bear" Christie. Several years later, by then approaching his fiftieth year, Christie lied about his age and enlisted in the Canadian Expeditionary Force. Serving with the Princess Patricia's Canadian Light Infantry, he was decorated twice for bravery in battle, and rose through the ranks to become an officer. Christie outlived his grizzly bear by thirty years.[82]

After Christie filed the "King Edward" claim on June 3, 1907, he had immediately transferred it to Harry McWhorter and William Williams.[83] The same year, Jack Stewart and Dr. William Catto staked the "Victoria" claim on a major vein containing gold and silver on the north face of Potato Hills Ridge, overlooking Dublin Gulch. These partners would continue to develop this property for many years. The following year, Bobbie Fisher staked the "Olive" mineral claim, which he named after his niece Olive Powers Kinsey. Fisher sold the claim to his sister Agnes Jane Kinsey in 1910, and she remained involved in the property until at least 1916.[84]

Bobbie Fisher came to the Yukon from Newfoundland during the Klondike gold rush. He mined on Dominion Creek in the Klondike district for several years, before moving to Mayo in 1906. He remained in this vicinity for the rest of his life. Fisher was a hard-working, optimistic prospector who found many properties for others, but never struck it rich himself.[85] Some years later, Fisher described how he became involved on Dublin Gulch: "It was in the year 1907 that I first went up to the Mayo district. I had the intention of prospecting for gold on the Potato Hills, at the head of Dublin Gulch. In 1908, after I had been prospecting around quite a bit, I found a quartz lead running northeast. This lead was about two miles from the Potato Hills and was seven feet below the surface of the ground. The rock of which it was composed was a greenish-blue quartz, filled with sulphur and arsenic, and carrying very fine gold.

"The formation of the hill on which the Olive claim is situated is made up of faulty granite, schist, pegmatite dykes and decomposed rocks. In these contacts I have found twelve leads. All of them run northeast and range from two inches to four feet in width. Some of them carry small quantities of molybdenite. But the best of all these leads is the one on the Olive and this claim is greatly valued by my sister, Mrs. A.J. Kinsey, who is present owner of it, or should I say two-thirds of it. She bought the claim in the fall of 1910 and had it Crown-granted in 1915–1916."[86]

William Aldcroft relocated the King Edward claim in 1908 and shortly thereafter sold it to Martin Joseph Raney and Henrietta Williamson, who in turn sold their half-shares to Jack Stewart and Dr. William Catto. Stewart and Catto formed a partnership on December 18, 1909, and remained active on Dublin Gulch for more than a decade. Dr. Catto was a Dawson City physician with an active interest in promoting and developing hardrock properties. He is most well-known for his involvement in the development of the Lone Star mine on Victoria Gulch, a tributary of Bonanza Creek, about fifteen miles from Dawson City.[87] Other Dublin Gulch claims owned by Stewart and Catto included the "Dublin King," the "Foundation," the "Happy Jack" and the "Victoria." Over the following years, they continued to develop their properties, with Stewart doing the physical work on-site while Catto did the promotion. By 1921, in addition to numerous surface cuts and shafts, Stewart had excavated 1,800 feet of tunnels.[88] Another active hardrock developer was Frank Carscallen. In 1909, he staked the "Margaret" claim and grouped it with the

"Mexican," the "Ophir," the "Midas," the "Blue Grouse" and the "Sore Leg" claims. Carscallen practised law in Napanee, Ontario, before coming to the Yukon in 1897. He lived in the Mayo district for thirty years and served a term on territorial council from 1928 to 1931, representing the Mayo district.[89]

The outlook was optimistic for the hardrock prospects in 1910 in the Dublin Gulch area and the Yukon in general. It was estimated by an unidentified source in the *Dawson Daily News* that roughly a quarter of a million dollars had been invested in developing hardrock properties, including twenty-five thousand dollars on the Lone Star deposit. The article stated that there was "work on the Dublin properties, where Stewart and Catto have a group with tunnel far in; and other work on Dublin now being pushed by Wooler and Fisher." The article noted that most of this investment came from people living in the Klondike: "They are showing their faith by reinvestment."[90]

The work continued in 1911: Stewart had driven his tunnel in another eighteen feet, though he did not state on which claim the tunnel was located. The quartz vein he was working, though a mere four inches wide on the surface, had quickly expanded to eighteen inches where Stewart had tunnelled.[91] Meanwhile, Bobbie Fisher was reporting optimistic results for the Olive mine. The *Dawson Daily News* stated that "on the Olive, the rock shows up so well the gold easily can be washed out with a pan. Ledges and stringers are running in every direction... Although Fisher got it first, Stewart also has it on his ground, and Cascallen, half a mile further up, has the same rich vein. Cascallen can pound out gold the same as the others. On the Olive, Fisher has a solid five-foot ledge. He got the best nearly at the top of the mountain, and three hundred feet below he got it on the creek."[92]

It was looking so favourable for hardrock prospecting on Dublin Gulch that men were reported to be quitting their jobs so that they could work on the hardrock propositions. An editorial in the *Dawson Daily News* in September 1911 stated, "A movement is on the foot to place a mill or two on Dublin Gulch, on the upper Stewart, tributary to Dawson." It further rhapsodized, "This widespread activity in the stamp mill work in the North is exhilarating... The new era is dawning in the North. Let everyone encourage the miner and by rational and sure process work toward the realization of the great destiny of the Northland."[93]

Returning from Dublin Gulch in 1911, sourdough Klondiker Curley Munroe reported that Dublin Gulch was the biggest thing he had ever seen, as described in the *Dawson Daily News*: "The property owned by Fisher and Sprague is where the rich pay has been struck. Two men who went up the last trip with us took a large steel mortar device for crushing rock, and it was reported they were pounding out forty dollars a day on a lay.

Curley had gone over the ground and found no quartz open for staking closer than eight miles from the original property. He wanted to buy, but found the property held at impossible figures for him... There is talk of sending a stamp mill or two to the property this fall. Mr. Sprague will not decide, he says, for a few days what will be the plan for the operations the coming winter."[94]

There was no more vocal mining promoter in 1911 than Bobbie Fisher. He was in Dawson City for Christmas 1911, and dropped by the Dawson News office to show off samples of the rock, even leaving some for others to examine. Fisher was close-mouthed about his own property but expressed confidence that it would prove to be of great value. He sank a shaft down thirty feet, and the vein he was following was more than three shovel-lengths in width. Five feet of the vein, he said, was exceedingly rich. Assays showed more than three hundred dollars to the ton.[95]

The Olive adjoined the Sprague group, which in turn was shoulder–to-shoulder with the Stewart and Catto property. Fisher was confident that all three properties would prove their worth. Despite lacklustre reviews from visiting experts, Fisher said that investors were anxious to learn all about the work. Bowles Colgate Sprague had staked hardrock claims on Dublin Gulch the year before. His Blue Lead group consisted of eight claims that contained a six-foot vein that yielded $2.50 to $27 per ton for several hundred feet. Sprague was a veteran of the American Civil War who held large silver interests on Keno Hill, and a farm in Sunnydale across the Yukon River from Dawson, as well as the property at Dublin Gulch. Despite his advanced age, he married his second wife, Annie Hall of Los Angeles, in a quiet ceremony in Dawson City on Christmas Day, 1911.[96]

Quartz claims had been staked throughout the Yukon in the decade following the Klondike gold rush, primarily in the Dawson City area, but also in the developing properties in the Mayo district. With placer production and the general population both declining, more attention was being paid to the hardrock opportunities that might be found in the territory. R.G. McConnell of the Geological Survey of Canada noted that plenty of quartz claims had been staked, but little progress had been made. Assay values for some of the sampled rock looked promising, but the paying veins of minerals in the Yukon were few and far between. On February 12, 1912, the Yukon Miners' Association appealed to the federal government to assist in the development of lode mining.

Two weeks later, Dr. Cairnes of the Geological Survey responded positively: "For a number of years past and particularly since 1905 when the government mill and assay office in Dawson were closed, relatively very few assays of the quartz of the Klondike

have been made... new deposits are being discovered each year, but little is known concerning the relative value of the bulk of quartz now known to occur in the district."

Thomas Archibald MacLean, a mining engineer from Nova Scotia, was hired by the Geological Survey of Canada to visit the Yukon and report on the quality and distribution of mineral prospects in the region. He hired his father-in-law, Dugald MacLachlan, to assist. Originally from Charlottetown, PEI, MacLean graduated from McGill University in 1898 with a B.Sc. in mining engineering. MacLean married MacLachlan's daughter in 1904, and became managing director of Bras d'Or Lime in 1907. He asked MacLachlan to accompany him to the Yukon because of his "experience in prospecting, sampling and milling in Nova Scotia gold mines."[97]

Dugald MacLachlan was born in West Bay on the shores of the Bras d'Or Lake on Cape Breton, Nova Scotia. By the time he was twenty-five, he was a well-established lime manufacturer and was operating two general stores farther south, in Hants County. After he married, most of his adult life was spent a little farther north of West Bay, where he developed a couple of mining enterprises as well as running the general store in what was to become the town of Marble Mountain. In 1886, he established the Bras d'Or Lime Company and the Bras d'Or Marble Company and was the first manager of each.

Dugald MacLachlan is the man with a white beard seated beside the entrance to the Olive mine.

The MacLean party en route to Dublin Gulch, 1912. The lack of a road forced the miners to haul in equipment and supplies in the winter when the ground was frozen. In the summer, transport was limited to walking or pack horse through spongy muskeg.

During the course of running two mining companies, he routinely visited mining operations throughout both Canada and the US to broaden his knowledge of mining.

The MacLean party spent June, July and early August of 1912 examining various quartz properties in the area surrounding Dawson City. On August 21, they boarded the steamer *Vidette* and were transported to Mayo three days later. They set out immediately for Dublin Gulch. For eight days, they explored various mining properties including the Stewart and Catto group, the Olive group, the Blue Lead group and the Eagle group. The first ten miles to Dublin Gulch were over a good wagon road as far as Minto Bridge, followed by another ten miles of rough wagon road to Look-out Cabin at the foot of Mount Haldane. Beyond that, the trail passed through muskeg and marsh, and was only passable during the summer months, by pack horse. "It will thus be seen," reported MacLean, "that the prospectors at Dublin Gulch are labouring under a severe handicap through lack of a good road for ingress and egress."[98]

MacLean found that the property of Stewart and Catto, a grouping of five claims located on the divide between Stewart and Olive pups, was the most extensively worked. Tunnelling and other work had been confined to two of these claims: Happy Jack and Victoria. Jack Stewart had excavated 128 feet of tunnel, but had not followed the vein into the hillside. Work observed at the Olive group of claims included "several surface trenches, a short cross-cut tunnel and drift, the work being done by J.E. Moskelund, under a lay agreement." Samples were taken for assay, and MacLean concluded that the "property evidently warrants further prospecting."[99]

B.C. Sprague's Blue Lead grouping of claims was promising, according to MacLean. A twenty-five-foot shaft had been sunk on Stewart Pup. Several surface trenches on the property suggested "the probability of a well-defined vein being uncovered by further prospecting."[100] The Eagle group, held by Sprague and others, was also examined, and samples were taken for assay from various exposures and open cuts. The assay results were favourable, and further prospecting was recommended.

Geological party at the Look-out Cabin on the trail to Dublin Gulch.

MacLean compared the results of his work on Dublin Gulch with findings at the Lone Star deposit on Victoria Gulch, concluding that the Dublin ore would have to contain at least twice the amount of gold to make mining viable. Three reasons were cited for this: the nature of the deposits, the distance and difficulty of access, and the greater cost of recovery. MacLean concluded, optimistically, that although the current values of the samples assayed fell below the level that would permit economic mining of the properties, "there is a strong possibility that further development, accompanied by more detailed work than was on this occasion possible, might result in establishing beyond reasonable doubt, the existence of one or two good mines."[101] Having concluded their examination of the Dublin Gulch prospect, MacLean and MacLachlan returned to Mayo and paddled down the Stewart River to the Yukon River and Dawson City, which they reached on September 13, 1912.

Harry McWhorter, an American prospector who had staked claims on Dublin Gulch in earlier days, had moved on to Alaska, where he made a pile of money selling a claim in the Iditarod region to the Guggenheims. He returned to the Yukon in 1912 and in the

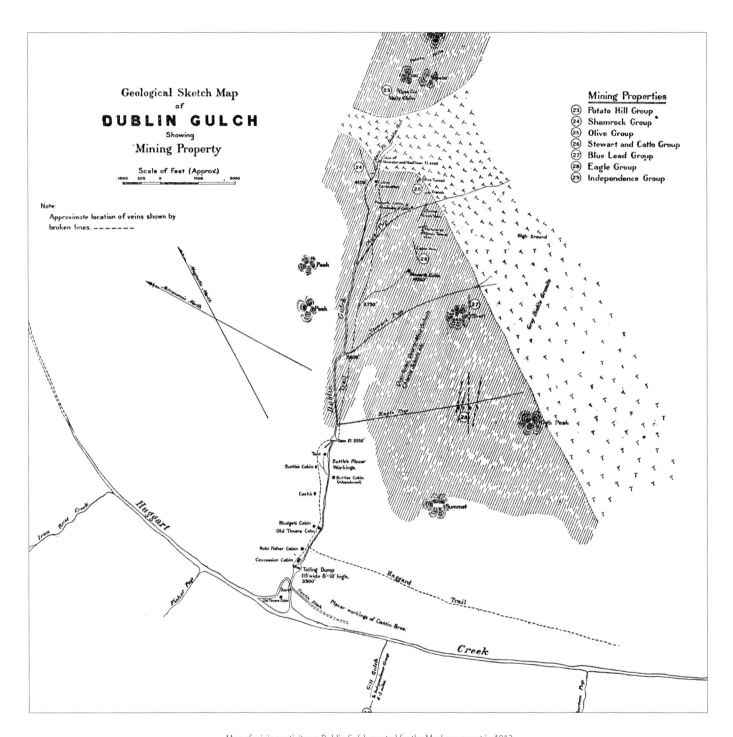

Map of mining activity on Dublin Gulch created for the MacLean report in 1912.

fall he partnered with Jack Alverson, a veteran Oregon miner, and they took a lay on the Midas and Ophir claims. Alverson had a cabin at the upper end of Dublin Gulch that he shared with Grant Huffman, a Scotsman who had been in the Mayo district for more than a decade. In February, McWhorter purchased the Midas claim from Frank Carscallen for fifty thousand dollars (with a five-thousand-dollar down payment), and then the duo worked their way up the McQuesten River valley.[102]

They met Grant Huffman and Mark Evans, and together the four men continued to a property on what later became known as Galena Creek. This claim had been abandoned some years before by Jake Davidson, and samples he gave to McWhorter from this property assayed at three hundred ounces of silver to the ton, but McWhorter was looking for gold and paid no heed to it at the time. They arrived at Davidson's abandoned property and McWhorter restaked it as the "Silver King." In the spring of 1913, he gave Alverson and Huffman a 100 percent lay on the claim in exchange for building a cabin on the property and doing some development work. Within a year, Alverson and Huffman had stockpiled sixty tons of silver ore, which was shipped out to the Selby smelter in San Francisco in the summer of 1914. When the shipment yielded nearly sixteen thousand dollars, it caused a staking rush and soon forty men were locating claims in this area.

Alverson and Huffman were so successful with their work in the Silver King that McWhorter cancelled his lease arrangement with them. Two years later, Thomas Aitken purchased the Silver King from McWhorter for seventy-five thousand dollars, and high-graded the property. By the time he was finished, he had earned nearly a half million dollars, and for the next half century, mines in this vicinity continued to yield high-grade silver ore and drive the regional economy. McWhorter later returned to Alaska, where he made another fortune. He and his wife left Alaska and settled in California, where he bought a cattle ranch and financed a bank.[103]

By 1914, there were men prospecting and mining on all the creeks and tributaries of the Mayo district. Mining activity on Dublin Gulch continued on the major groupings of claims. Again, mining on the Stewart and Catto group was focused upon two claims: Happy Jack and Victoria. In one, work had progressed two hundred feet into the hillside with a one-hundred-foot crosscut, while the other, higher up the hillside, ran in 1,275 feet with seventy-foot crosscuts, exposing a vein that assayed from four to fourteen dollars per ton. The adjacent Olive claim had a tunnel running one hundred feet into the granite with samples yielding nine to fifty dollars per ton. Meanwhile on Frank Carscallen's Midas claim, two tunnels—one of 35 feet and the other of 150—produced samples running nine to thirty dollars per ton in a seven-foot-wide vein. Sprague's Blue Lead was returning values of $2.50 to $30 per ton for several hundred feet; R.S. Ames, who had dug a twenty-seven-foot shaft on this claim, was preparing to start drifting.[104] Meanwhile, Sprague's Eagle group yielded samples from sixteen to twenty-seven dollars per ton.

All the excitement and growing interest in mineral potential did much to stimulate the economy.[105] A *Dawson Daily News* editorial in February of 1914 acknowledged the increasing mineral development around Mayo and optimistically forecasted a stable future for mining. It went on to state that the introduction of a diamond drill would advance the district "by leaps."[106] The town of Mayo started to grow: the mining recorder was relocated from Minto Bridge to Mayo, and a school was established. Jim Ferrell and his wife, who had been trading at Lansing Post farther up the Stewart River, moved to Mayo and opened a store. New businesses included a blacksmith, a sawmill, a livery stable and a liquor store. River traffic had increased to the point where there were regular weekly steamer trips between Dawson City and Mayo, and during the winter months there was bimonthly stage service between Dawson and Minto Bridge.[107]

War was declared between Britain and Germany on August 4, 1914, and over the next four years, it drew many men away to fight in the Canadian Expeditionary Force overseas. Meanwhile, Dublin Gulch continued to be a promising mining location during the years of conflict. In 1915, all the miners working there seemed to be highly pleased with their workings and were optimistic that Dublin Gulch would become a great quartz camp.[108] Among those active in 1915 were Bobbie Fisher, George Graham, John O'Brien, Jack Stewart (of Stewart and Catto), Lloyd Trombley and William Walby.[109] After examining the area, Dr. Cairnes of the Geological Survey of Canada "reported considerable scheelite tungsten on Dublin gulch, and said the proposition looked promising."[110]

Early on, the placer miners on Dublin Gulch noticed a heavy white sand in their sluice boxes, which geologist Joseph Keele of the Geological Survey of Canada later identified as scheelite. This product was nothing more than a nuisance to the placer miners, and no further interest was shown to the mineral until 1912, when another geologist from the Geological Survey encouraged the miners to save this light-coloured by-product. They subsequently started putting this mineral aside at the rate of 182 kilograms per week. In 1916, Bobbie Fisher discovered scheelite-bearing quartz veins near the head of Dublin Gulch.[111]

Scheelite had strategic importance to the war effort and interest in the region grew as a result. Dr. Alfred Thompson, the Yukon's member of parliament, lobbied hard to get a government purchasing agent located in Dawson to buy tungsten ore for the war effort.[112] His efforts came to nothing. Bobbie Fisher explored the area of the junction of Olive and Dublin Gulches, locating several small lode deposits of scheelite, but he did little or no work on these prospects.[113] The *Dawson Daily News* editorialized on the matter, stating two things that were holding back the development of the tungsten deposits: lack of expert advice on how to proceed, and poor transportation to get the ore to the refinery.[114]

Yet the diehard prospectors of Dublin Gulch persisted. By 1921, Bobbie Fisher had extended the tunnel on his claim to two hundred feet, "in which he found gold-bearing ore of much assay value." Meanwhile Stewart, the most industrious of all Dublin Gulch miners, had accumulated a total of 1,800 feet of tunnels, and remained abundantly confident of the future of the property.[115]

But interest in Dublin Gulch was eclipsed by new discoveries farther up the McQuesten valley. In 1918 Louis Bouvette, one of a number of prospectors staking around McWhorter's Silver King mine, found rich galena float on the north side of Keno Hill. He followed the float uphill until he found a frost-shattered vein and staked it the following year. Samples were taken to Dawson City, where A.K. Schellinger of the Yukon Gold Company assayed them. The results (two hundred to three hundred ounces of silver per ton) were enough to entice Schellinger to the area to investigate. Yukon Gold acquired a number of claims on Keno Hill, including Bouvette's original "Roulette" claim.

A staking rush followed and hundreds of additional claims were added to the books in the following year. Yukon Gold formed a subsidiary called Keno Hill Ltd., and sank its first shaft on what became known as the Sadie-Friendship vein. A warehouse was built in Mayo and waterfront was leased for stockpiling ore, the first of which was shipped out from the mine in the winter of 1920–21.

The year 1921 was probably a high point in the history of Keno City, both for the active growth in the community and in the optimism for its future. In addition to Keno Hill Ltd., Treadwell Yukon Company was established there that year, spending a half-million dollars on options on various mining properties. The population of the Yukon had declined from eight thousand in 1911 to half that number by 1921. The population of the Mayo district was estimated to be slightly less than one thousand individuals, almost a quarter of the population of the entire territory. Liquor sales from Keno City and Mayo accounted for one-quarter of the total revenue for the territory. In 1921, to supply the demand for meat in the newly established town, T.C. Richards, the manager of the P. Burns & Co. store in Whitehorse, drove a herd of thirty cattle overland to Keno City. From Pelly, where they had been unloaded from the steamer *Whitehorse*, they travelled overland to Mayo, and then on to Keno City, where they were slaughtered and sold to various hungry customers.[116]

The Mayo district had become important enough economically that it demanded the attention of the politicians. Though there was not a separate riding for Mayo until the territorial election of 1928, there were enough voters now located in this district that candidates in the 1921 federal election couldn't ignore the region. There were three candidates—George Black, Frederick Congdon and George Pitts—but only the former two were serious contenders.

Black campaigned hard for votes, having trekked overland on snowshoes through wilderness in cold winter weather from Fort Selkirk. He arrived at Mayo on November 7, and visited the various communities in the district. Black held public meetings at both Mayo and Keno City, and at both large mining camps on Keno Hill. His final speech was delivered to 150 people at Mayo, after which there was dancing and refreshments.[117] Black's efforts paid off—he took seven of the eight polling stations in the Mayo district and beat Congdon 145 votes to 123, with Pitts a distant third.[118] There was a polling station at William Abbott's cabin on Haggart Creek, where the vote went overwhelmingly to Black, eight votes to one.[119] Black carried the seat for the Conservatives, though the Liberals, under William Lyon Mackenzie King, took the nation.

Another indication of the importance of the region came the following year when the Governor General of Canada, Lord Byng of Vimy, visited the Yukon. While his wife remained in Dawson City, the Governor General and a herd of local officials visited the Mayo district. His Excellency toured the mines at Keno Hill, where he met and shook hands with practically everyone in town. The weather was ideal for the journey, although the roads were not, despite extensive work to upgrade them. According to George Black, "The road was very rough in spots and a short distance below Keno City, his car went down to the hubs in the mud and became hopelessly mired. When the other cars in the party came up His Excellency was hard at work helping to shovel the way out. By combined efforts of the party the car was pulled out and got under way again."[120] No mention was made of Dublin Gulch, which lay isolated and sparsely populated.

The focus of mining in the district had shifted to the silver mines around Keno Hill. Although not extinguished entirely, prospecting in the Dublin Gulch area continued on a reduced basis during the 1920s. Newspaper reports continued to make reference to the potential of the Dublin Gulch area, qualified by concerns that the difficulties in transportation to the creek were retarding development.[121] Jack Stewart and Dr. Catto continued to develop their group of claims on the gold-bearing vein, and the prospects looked promising.

Other claims in the vicinity were reported as having leads yielding better assay results than those in the Stewart and Catto group. On a number of these claims, considerable high-grade scheelite, both in placer deposits and in hardrock veins, was found. The *Dawson Daily News*, reporting on this, added that several tons of tungsten concentrate were shipped outside during the war.[122] Bobbie Fisher continued to bring optimistic reports with him to Dawson City whenever he visited. In 1921, he reported assay values of such quality that he was "satisfied it will be worked in time."[123]

Fisher was also confident that the ground being developed by Stewart and Catto would become a steady producer.[124] The following year, Fisher was back in Dawson City, this time reporting values so rich as to stretch his credibility to the limit. He said that some assay samples were reporting values of ten thousand ounces to the ton, with others in his Olive group ranging in the hundreds of dollars to the ton. "The veins are wide," he reported, "and the two hundred feet of tunnel on my property and the hundreds of feet of tunnel work done on other properties in that locality are but preliminary, I feel sure, to greater activity in that district."[125] Robert Henderson, the Canadian government's officially designated discoverer of the Klondike, came back from the Mayo area that fall also bearing an optimistic outlook.[126]

B.C. Sprague also carried favourable news about the potential for Dublin Gulch when he returned from a trip to the area in 1919, but it was not gold he was crooning about; it was tungsten. It was noted by the *Dawson Daily News* that Dr. Cairnes had visited the Gulch some years before and stated that this was the most favourable deposit he had seen in the Yukon. The ore was found in several parallel veins running up the left limit of Dublin Gulch, in one place a vein three feet across being reported. These remarks were sufficient to sustain the continued activity of Stewart and Catto in this area of the Yukon.[127]

But Sprague would not continue to be a player in the development of Dublin Gulch. Within a couple of years, he was forced to leave the territory due to ill health. He first moved to an American soldiers' home in LA, then later to one in Virginia, where he died intestate on August 25, 1925. At the time of his death, he owned a large house in Dawson and numerous mining properties in the Mayo area.[128]

During the 1920s and 1930s, interest in Dublin Gulch dwindled but did not die entirely. A geological report from 1928 noted veins of scheelite returning values of 10 percent tungsten.[129] Bobbie Fisher held on to his Dublin Gulch property until his death in 1939, ever optimistic of the promise his properties held for future mining.[130] In 1937, Tom McKay and Archie Martin staked claims on Dublin Gulch next to the Olive group. They prospected with open cuts and shallow shafts in 1938, and sold the property to Treadwell Yukon Company. Treadwell did considerable exploration work on the property in 1941 under the direction of A.K. Schellinger, but the work was abandoned when the results proved to be unsatisfactory.[131] The property was later transferred to Keno Hill Mining Company in 1946.[132] It was restaked in 1948 as "Avoca" by J.J. Colt and J.B. O'Neill, who trenched by hand and with machinery from 1949 to 1954. They sold a share in the property

Geological Survey pack train on the Haggart Creek road. To bring tungsten ore out from Dublin Gulch for the war effort, a road was extended from the McQuesten River to Dublin Gulch in the summer of 1942.

in 1958 to Ed Barker, who trenched the property for three years before he, in turn, sold the property to Peso Silver Mines Ltd., which did further trenching in 1962.[133]

By 1955, Colt and O'Neill held a block of fifty-six quartz claims covering an area of Dublin Gulch that included patented claims that were originally staked in 1903. Sufficient work was done to obtain the patents only on these early claims, and that pattern continued in 1956 and 1957, when they allowed fifteen of the claims to lapse. Eventually, the remainder of the claims became open and were restaked.[134]

Nearby, prospector Harvey Ray staked the "Tip Top" claim on an exposure of scheelite-bearing float in 1942. The source of the float was located the following year by a member of a Geological Survey party examining the area.[135] Because it was wartime, the government took a special interest in strategic minerals. Consequently, the Geological Survey of Canada investigated the property from 1942 to 1944.[136] The government held high hopes for developing tungsten properties, so in 1942, twenty thousand dollars was allocated to finance the construction of a road to Dublin Gulch.[137]

Roadwork began in the spring. The bridge over the McQuesten River, with a "forty-foot centre span," was completed before breakup in the spring. The road, which was well graded and well drained, reached Dublin Gulch by midsummer. The total cost for the roadwork including the bridge was seventeen thousand dollars. Neil Keobke, the

Left to right: Alec Berry, Ed Barker and Kay Broadfoot in front of Barker's cabin on Haggart Creek.

road foreman, chose the route of the road, designed the bridge and oversaw the work. The balance of the money allocated for the project was used to repair equipment for work the following year.[138] The gulch attracted wartime interest because of the increased demand for tungsten, and several prospectors staked, hoping to take advantage of the opportunity. A considerable amount of placer scheelite was recovered in "satisfying quantity by several small operators," some by re-mining their tailings, but no major hardrock development occurred during the war.[139]

After the war, Dublin Gulch began attracting more attention from corporate mining interests. The Ray Gulch claims were restaked in 1951 by Ruth Batty and Ed Barker. Stride Exploration and Development Company prospected and sampled the exposures in 1956. Ed Barker and associates held fifty claims opposite the mouth of Dublin Gulch, on which they had done considerable surface work with bulldozers. The properties covered ground that contained gold, scheelite and tin prospects.

In the words of geologist Aaro Aho, "Hardrock prospecting became Ed Barker's main interest in later years and he worked at it and talked about it continually. Right beside his cabin and up the creek, placer mining had exposed several veins carrying antimony, lead and minor silver, copper and zinc. These veins, the gold veins up nearby Dublin Gulch, the tin on Tin Dome, the scheelite found by Bobbie Fisher in 1917 at the head of Dublin Gulch, and the fortunes that were made on the silver in the Mayo district had caught his imagination. By doing assessment work with his bulldozer on the claims he staked, he gradually acquired control of the large group staked by John O'Neill in the 1950s, and for years he tried to get geologists or engineers interested in spending time in the area, in taking over some of the large groups of claims in which he had an interest or in giving him technical guidance. However, no one did more than come in for a few days to look at the showings he had exposed, none of which proved promising enough."[140]

Things started to heat up during the 1960s. In 1960, Mayo Silver Mines Ltd. located a promising silver vein on the east side of the headwaters of nearby Ray Gulch, but did not explore the property for tungsten.[141] The area was also examined by United Keno Hill Mines Ltd., which prospected, sampled and trenched, but the results weren't satisfactory and the claims were allowed to lapse. In 1968, a block of 116 contiguous claims was restaked by Conrad Provencher and John Boyce. In the middle of this property was the original Olive mine, still owned by the heirs of the original owners. Provencher optioned ground to Great Plains Development Company of Canada, Ltd. in 1968, Tam Mines, Ltd., in 1969 and Connaught Mines Ltd. from 1969 to 1971. In 1969, Connaught subleased the Mar-Tungsten zone to Canex Aerial Exploration Ltd., but returned it in 1971. Great Plains undertook bulldozer trenching in 1968 and samples were collected, but no further exploratory work was undertaken before 1970.[142]

In 1971, Canex Placer drilled three holes and dug twenty trenches to evaluate the low-grade quartz-scheelite vein for tungsten and took soil samples for geochemical testing from the area overlooking the north side of Potato Hills Ridge above Dublin Gulch. Six years later, under the corporate umbrella of the newly formed Queenstake Resources Ltd., Gordon Gutrath staked thirty contiguous claims in the vicinity of Ray Gulch, and conducted a "small program of geological mapping of the skarn zones near Ray Gulch."[143] In the fall of 1978, after having spent the summer placer mining on Dublin Gulch, Gutrath met with the president of Canada Tungsten. The result was a partnership, with Canada Tungsten purchasing a 30 percent share of Queenstake. Canada Tungsten also staked adjacent claims and purchased those owned by Ron Holway and Darrell Duensing.

View from Potato Hills of the Mar or Ray Gulch tungsten deposit in 1979. The trench (cat road) goes through the centre of the deposit and the road demarcates the length of the deposit.

Janet Dickson was thirty years old when she met Yukon prospector Gordon Dickson in Vancouver. She first came to the Yukon with him in 1967 after they were married. She didn't know the prospecting game when she started, but she took to it like a duck to water. She liked the out-of-doors, and she learned the business quickly. Thereafter, she looked for minerals all over the Yukon Territory: Watson Lake, Bonnet Plume, Mount Nansen—and Dublin Gulch. The Dicksons continued their prospecting partnership until Gordon passed away in March 1993, after which Janet continued on her own until she retired.

Unlike placer mining, quartz mining is a rich man's game, but you can make a living at prospecting if you play it right. When Janet joined the game, they could only stake eight quartz claims within a given area, so if they wanted to stake a large tract of land, they would hire others to stake on their behalf. The rules on this have changed, but that is the way it was when she started. The Dicksons would find attractive deposits, stake them and then try to interest exploration companies to come and test the ground for mineral potential. The claims were optioned to the company under specific terms and conditions, and they drilled or trenched and employed other specialized technology to expose the mineral values of the property. If they liked what they got, they could purchase the claims outright; if not, they reverted to the prospector.

When Gordon and Janet Dickson staked claims on Dublin Gulch in 1975, they were looking for tin. The market price of tin at the time had gone up and Gordon knew of a showing on Dublin Gulch. So they set up their tent in the valley below, and every day would climb the hillside to do the staking. "In those days I was a smoker," Janet recalls. "I used to count the steps up the hill—five hundred every morning, and we uncovered a cassiterite vein. I never knew they looked like diamonds. [To me] it looked just like a diamond mine."

They worked it by hand at first, and then they hired Ron Holway and Darrell Duensing to bring up a front-end loader to do the assessment work. Holway and Duensing came in with their loader, and with one scoop of the loader bucket, Janet's "diamond mine" was gone. The Dicksons worked the property for three years and then optioned the claims to Canada Tungsten in 1978. Can Tung was exploring for tungsten. The source was found in the first hole drilled in 1979 and a large tungsten skarn deposit was outlined by approximately forty thousand feet of drilling. This deposit, known as Mar-Tungsten, was located at the headwaters of Dublin Gulch and Ray Gulch.

According to geologist Brian Lennan, "Can Tung planned on driving a development adit into the heart of the deposit in 1981 to test mining conditions; however, the Chinese government dumped tungsten on the market and drove the price of scheelite down from $150 for a short ton unit (twenty pounds) to $45 per short ton unit. This forced the closure of the Can Tung mine in the Northwest Territories and the shelving of driving the adit into the Dublin Gulch Mar-Tungsten deposit."[144] Can Tung held the option until 1986 and ran a large placer operation on the Dublin Gulch property for a number of years.

The Dicksons also staked a large number of quartz claims nearby on behalf of Can Tung as part of the deal, with the understanding that if the company ever abandoned the claims, the property would revert back to the Dicksons. Can Tung abandoned Dublin Gulch in 1986, and the Dicksons then optioned the ground to Queenstake Resources under terms and conditions that continued to apply to every successor to Queenstake. Included among the Dickson claims were the "Smoky" claims that sat directly upon the Eagle zone, which was to later take on great importance. The Smoky claims were named after Smoky Sheldon, one of the people hired by the Dicksons to stake claims that blanketed the hillside overlooking Dublin Gulch. Adjacent claims were staked in the names of others hired by the Dicksons to do the job: "Jeff," "Dave" and so on. Looking at a map of the Dublin Gulch hardrock claims, Janet Dickson says, "You prospect all your life, you know, but if you hit it lucky once… that's all it takes." And lucky she was, as the future Eagle Gold Mine was situated on claims held by the Dicksons. But as important as the claims you stake are the deals you make.

Canada Tungsten undertook the first large-scale hardrock exploration of Dublin Gulch. In addition to the placer mine they operated for several years, the company retained Bema Industries to manage the exploration work on property consisting of 1,059 full or fractional quartz claims overlying the area. The first work that Bema undertook consisted of nine kilometres of bulldozer trenching, with samples running as high as 5 percent tungsten per ton. During 1979 and 1980, nearly fourteen kilometres of cores were drilled, followed in 1982 by three additional test holes for a total of 750 metres. Based upon the results of this work, Bema recommended further drilling.[145] Instead, Canada Tungsten returned the claims to Queenstake Resources, with the exception of the Smoky and Bob claims, which were returned to the Dicksons.

From 1987 onward, further work on Dublin Gulch focused upon the hardrock gold that existed there. After Can Tung returned what was known as the Mar-Tungsten zone and adjacent gold claim blocks, Queenstake subsequently drilled four core holes on two gold veins on the property for a total of 705 metres.[146] In 1988, more drilling on Dublin Gulch returned assays of up to 11.2 grams per tonne. A chip sample from veins located on the bottom of Dublin Gulch yielded values of up to forty-one grams per tonne. The property was optioned to Can Pro Development Ltd. in 1989. Trenching on the Smoky 64, R&D 6, R&D 16 and Bob 3 claims in 1989 between Olive and Stewart gulches exposed three new vein systems returning values up to 8.61 grams per tonne of gold.[147]

By 1991, Dublin Gulch was receiving much interest from the corporate world. The property was acquired by Ivanhoe Goldfields Ltd. that year. In 1992, it was optioned to H-6000 Holdings/Amax Gold, and attention on the property shifted to exploring for veins similar to the mineralization also found at the Fort Knox gold deposit forty-two kilometres from Fairbanks, Alaska.[148] Keith Byram of Pelly Construction recalls that his company was hired to run a D9 Cat on the property that year, digging test holes.

The Fort Knox gold in Alaska is found in an intrusive late-Cretaceous granite pluton. Gold occurs along margins of quartz veins, on quartz-filled shear zones and along fractures within the granite. Fort Knox operates as an open-pit mine. The richer ore is milled, while the gold in the lower-grade ore is recovered by the heap leach process.

Placer gold was first discovered in the Tanana district of Alaska in 1902, and a stampede ensued that resulted in the founding of the city of Fairbanks. The land where the Fort Knox mine is located was first staked in 1913 by H.A. Currier, but little development followed.[149] The ground was restaked for placer mining in 1980 by two prospectors, Joe Taylor and George Johnson, who worked the ground from 1980 until 1982. Several small companies explored the ground between 1987 and 1991. Denver-based Amax Gold purchased the Fort Knox property in 1992 in a two-for-one stock swap and created Fairbanks Gold Mining, Inc., to run the operation.[150] Fort Knox was owned by Fairbanks Gold (51 percent)—of which 37.9 percent was owned by Ivanhoe Capital Corporation—and Gilmore Gold (49 percent), an American company.[151]

As part of the deal, Amax Gold obtained the option to purchase the Dublin Gulch property, but dropped it in 1992 in favour of other projects.[152] Construction of the Fort Knox mine started in 1995 and the first gold was poured the following year. Amax merged with Toronto-based Kinross Gold in 1998, and Kinross took ownership of the property.[153] The Fort Knox mine, which is already the top gold producer in Alaskan history, topped the eight-million-ounce plateau in 2019. The current life is projected to continue until 2030.[154]

Meanwhile, drilling on Dublin Gulch in 1992 focused on four zones of mineralization: Eagle, Olive, Shamrock and Steiner. The following year, the drilling zeroed in on one of these, Eagle, which appeared to be the largest gold-bearing zone of the four. In 1993, the property was returned to Ivanhoe Goldfields Ltd. (a Robert Friedland company), which had allocated $1.5 million for exploration.[155]

In July 1993, Ivanhoe conducted reverse-circulation drilling and backhoe trenching on Smoky claims 3 and 4 and fractional claim 96.[156] A year later, in February 1994, Starmin Mining, out of its Toronto office, acquired Ivanhoe Goldfields and Desarollos Minerals Ivanhoe Holdings, Ltd. In August, this new venture became First Dynasty.[157] First Dynasty, which was described as the Friedland family's fastest-growing investment, was headed by Timothy Haddon, who formerly headed up Amax Gold. Haddon met Robert Friedland in 1991, during the Amax takeover of Fairbanks Gold. Haddon left Amax in 1993 when Cyprus Minerals (Amax Gold's major shareholder) merged with Amax.[158]

The stakes were increasing each year. In 1995, First Dynasty raised nearly $13 million that was earmarked for Dublin Gulch and another property in Indonesia.[159] That year, First Dynasty drilled more than eight thousand metres on the Dublin Gulch property during its exploration season.[160] Optimistic estimates pegged the gold content at

3,740,000 ounces of gold in 110 million tons of ore.[161] Much had been done to develop the Dublin Gulch property. Between 1991 and 1996, more than two hundred holes totalling more than thirty-five kilometres had been drilled.[162]

First Dynasty exploration crew, 1995.

In 1996, First Dynasty reported that the Dublin Gulch property contained more than a million ounces of gold. That same year, First Dynasty formed a new subsidiary, New Millennium Mining Ltd., and transferred the Dublin Gulch property to it. New Millennium set about determining the extent and value of the Mar-Tungsten deposit for an open-pit mine. In 1996, they drilled six kilometres of additional holes in the Eagle deposit and established a reserve of fifty million tonnes, averaging 0.93 grams per tonne. Meanwhile, more exploration was done on the Potato Hills, 3.5 kilometres to the northeast, and fractional claims were staked around the Smoky claims to blanket the area under one owner.[163]

By now, New Millennium had pegged the reserves on the Dublin Gulch property at

Placer mining
areas during
the hardrock
exploration
program.

1.5 million ounces.[164] The company engaged consultants to produce a feasibility study, which was completed in 1997. This report projected that the property could be developed at a capital cost of US$107 million. The report stated the feasibility of a seasonal, heap leach operation capable of producing 135,000 ounces of gold per year at a cash cost of US$222 per ounce, with a strip ratio of less than one to one.[165] Unfortunately, the price of gold started to decline, and with it so did the interest to undertake further exploration of the New Millennium property.

In September 1997, First Dynasty was reported to be focusing on properties in Armenia and Asia. Another company, Cornucopia Resources, was planning to take over New Millennium (and the Dublin Gulch property). The deal would involve Cornucopia acquiring New Millennium, a wholly owned subsidiary of First Dynasty, for forty-five million of its shares. When the transaction was complete, First Dynasty would control a 53.9 percent equity stake in Cornucopia.

Over the period that New Millennium had the Dublin Gulch property, the company invested

Streamflow measurements, 1995.

more than US$10 million. The feasibility study had been completed, and New Millennium was in the final stages of the permitting process. Production was projected to begin as soon as 1999.[166] Unfortunately, the price of gold continued to tumble until it fell below three hundred dollars per ounce. In November, Cornucopia announced that the takeover was cancelled.

"In the process of our due diligence and in the process of this fall in the price of gold, it became apparent that it didn't make sense to each of the parties," said Dale Wallster, a representative of Cornucopia.[167] With a gold price down to $317 per ounce at the time, there wasn't much room for profit. The feasibility study was based on a gold price of four hundred dollars per ounce. First Dynasty, however, planned to complete the environmental permitting for Dublin Gulch and pursue an impact and benefits agreement with the First Nation of Na-Cho Nyäk Dun. The company intended to evaluate the feasibility of the project and, if conditions were favourable, advance it to the development stage.

After the collapse in the gold price, exploration of the Dublin Gulch property was

The northerly view of the Eagle Pup valley from the ore body. The Dublin Gulch placer mining area is in the background and the main access road to the ore body area is at the left.

The lower Dublin Gulch area that had been worked by placer mining. The proposed open pit area is on the ridge in the background.

Below: The first StrataGold camp at Dublin Gulch, 2003.

Exploration diamond drilling in the Eagle zone, June 2006.

placed on hold. There is no record of drilling or exploratory work between 1997 and 2004. The property remained with New Millennium until July 17, 2002, when First Dynasty changed its name to Sterlite Gold Ltd.[168] Two years later, in October 2004, Sterlite Gold was acquired by StrataGold Corporation, and StrataGold performed more exploration of the Eagle deposit. In 2006, Wardrop Engineering Inc. was able to estimate an "indicated resource of 66.5 million tonnes grading 0.92 g/t gold and an inferred resource of 14.4 million tonnes grading 0.80 g/t gold at a cutoff of 0.5 g/t gold."[169] Over the past twenty years, the claims overlying Dublin Gulch had been assembled under one corporate umbrella. The credit for that goes to Terry Tucker, who was the CEO of StrataGold.

Terry Tucker had spent a lot of time up in the Yukon. He was on the discovery team for the Selwyn deposit, and he knew Dublin Gulch. New Millennium/First Dynasty had the deposit in the mid-nineties and had completed a feasibility study and started down the permitting track. Then the gold price collapsed in 1997. The whole area got broken up and the momentum was lost. Terry Tucker was aware of it, and he got six or eight million dollars from Newmont Mining Corporation, and managed to get the different players and the package back together again. According to Mark Ayranto, a Victoria Gold executive, that is the truly great thing that StrataGold did.

By 2008, the Mar-Tungsten zone was just a small part of a larger holding that consisted of 1,896 claims, ten quartz leases and the Olive federal Crown grant, blanketing an area of 34,576 hectares. Slowly, over the previous thirty years, the value of the gold deposit on Dublin Gulch had been defined and evaluated. It took on the form of a three-dimensional body of rock with enough gold inserted within it to make it attractive for development.

The price of gold had risen from $273 per ounce in 2000 to $836 in 2007. Things were looking up. It was at this point that Victoria Gold entered the scene at Dublin Gulch, and a mine was about to be born. All it would take was enough mineable ore, the right technology, the right price of gold, the right team and the money to get the job done.

StrataGold CEO Terry Tucker (kneeling) and geologist Jim Sparling examining rock exposures in Dublin Gulch, August 2005.

CHAPTER 4

GROWING PAINS

——

Victoria Gold has origins dating back to 2002, but the current directors and management team became involved as a result of the $3.2 billion friendly acquisition of Bema Gold by Kinross Gold Corporation on February 28, 2007.[170] Founded in 1993, Kinross Gold is one of the largest gold producers in the world, with a diverse portfolio of mines and projects around the globe. With headquarters in Toronto, Kinross employs approximately nine thousand people worldwide.

Bema had a large interest in a company called Victoria Resources (which later became the Victoria Gold Corporation), which was exploring properties in Nevada.[171] The man who led the takeover and had to decide what to do with this orphaned company was Kinross executive vice-president Hugh Agro. Agro recruited Chad Williams, a professional engineer with an MBA, to be the chief executive officer. Among others who joined Hugh Agro on the board of directors were Sean Harvey and John McConnell.[172] McConnell had

The core shack, August 2011. Facing page: Aerial view of the one-hundred-person camp.

known Agro from previous business contact, but had never met Harvey before. McConnell and Harvey quickly became close friends. Over the years, the two men became partners in a number of business ventures, including a winery in British Columbia. When McConnell was married in 2018, Harvey was his best man.[173]

According to McConnell, "At the time Victoria was a small exploration outfit with a few early-stage projects in Nevada that, through various acquisitions, had become a subsidiary of Kinross Gold Corp., one of the world's largest gold miners. Victoria had little to show for itself at the time: no board, no cash, two million dollars in debt, and assets amounting to little more than a few good drill holes in the southwestern US." The first order of business for the little company was to find it new life. With financing from Kinross, they started looking around for potential acquisitions. Victoria obtained Gateway Gold in late 2008, which gave the company more assets in Nevada. At the same time, they turned their sights to StrataGold, the company that owned the Dublin Gulch claims.

Hugh Agro (in red hat), original Victoria Gold board member.

"I first looked at Eagle in probably 2007," says McConnell. "Again, it was Hugh Agro at Kinross. I think he saw something in Dublin Gulch that he likened a lot in similarities to Kinross's Fort Knox mine. So he suggested I have a look at it. I wasn't working full-time at that time, so I thought sure, it was interesting. And Kinross paid for me to go up to Fort Knox, look at Fort Knox and then go over to Dublin Gulch." McConnell liked what he saw. Both he and Agro then argued passionately to acquire the Dublin Gulch property.

According to McConnell, "It was drilled; there was a resource. I saw a lot of similarities to the Fort Knox deposit. It was open-pittable, it was big. My back-of-the-envelope calculations said, *There could be a mine here.* And it was decided, at Victoria, that we would approach then-owner StrataGold about buying the asset from them. We talked to them and they were just too far apart on what we felt the fair value was, so we walked away from it basically. Victoria continued to focus on the Nevada assets. And then along came the financial crisis in the fall of 2008 and as a company, Victoria saw that as an opportunity. Many CEOs were curled up in the fetal position underneath their desks. We approached Kinross and said we'd actually like to take advantage of this opportunity where many companies are badly beaten up, and we did a financing in the fall of 2008.

"In the fall of 2008 we acquired another company called Gateway Gold, which gave us more assets in Nevada. We also approached StrataGold, which held the Dublin Gulch property, and we only reached an agreement with them in early 2009, when they could not make the payroll. We made an offer to their shareholders that they found acceptable. I think we paid roughly ten million in Victoria paper, so no cash; we just exchanged Victoria stock for StrataGold stock."

As CEO of Victoria Gold at the time, Chad Williams was ecstatic about the acquisition: "I'm biased, but I've been in mining more than thirty years. I think it's one of the best deals I've ever seen... I think it's the deal of a lifetime."[174] Victoria Gold completed the acquisition of StrataGold Corporation on June 4, 2009. For the acquisition, Victoria issued twenty-three million common shares, worth $10.4 million, to shareholders of StrataGold.[175] The property had been consolidated, and the NI 43-101 resource estimate

John McConnell joined the board of Victoria Resources in 2007.

Historical placer mine workings at Dublin Gulch and Haggart Creek.

of 4.4 million ounces of gold had been acquired at a cost of only three dollars Canadian per ounce.[176] Kinross Gold was now the largest shareholder in Victoria Gold with 28 percent of the common shares.[177]

Paul Gray, the current vice-president of exploration for Victoria Gold, remembers the acquisition: "StrataGold did a good job. They blew up in 2008 when the market imploded, and Victoria came in and picked up the pieces. It was all timing... that's the business. StrataGold did a couple of good things. First, they kicked off the permitting process. That was when [Mark] Ayranto was there and [Hugh] Coyle, who's been around for twelve to thirteen years with the company—he's currently the longest standing employee with the company.

"They started the process. They did a good job with the community and with the water board, and they got all that going. And the most important thing, from my position, was that StrataGold consolidated the ownership of all the mineral claims and the placer leases—550 square kilometres. You can see it from space. It used to be fractional ownership—companies and individuals—and StrataGold pulled it all together into one cohesive package which had never been done before."

Victoria Gold now owned five core projects in the Yukon and Nevada, as well as a joint venture in Guyana with Newmont Corporation. To keep the ball rolling, Victoria issued more than four million dollars in a flow-through offering in order to finance the 2010 exploration season. In the meantime, McConnell made a quick inspection visit to Victoria Gold's Guyana property, and a decision was made to sell it and focus on the American and Canadian projects instead. In April, Chad Williams announced that Victoria Gold had sold the subsidiary that held its interest in the Guyanese project to Takara Resources Inc. for Takara shares, which would be disbursed gradually to Victoria over a three-year period. This left Victoria Gold with a 42 percent interest in Takara and two Victoria representatives, John McConnell and Marty Rendall, on its board of directors.

In 2010, Victoria Gold set its sights high and embarked on the long and sinuous path to developing a mine. The first step in that direction involved further exploration, permitting and feasibility studies. This is where Mark Ayranto played an important role. Ayranto hadn't planned to make his career in mining; he fell into it by chance. He took an undergraduate degree in marine biology at Dalhousie University, and went overseas to work in Indonesia with his wife for several years.

Ayranto's mentor there was the very first person to commercialize shrimp farming. They lived quite remotely in the era before internet, their communication link being

Mark Ayranto, Victoria Gold's chief operating officer.

single side-band radio to Surabaya, which was over three thousand kilometres away. It was a great experience for him, but there is no shrimp farming or aquaculture in Canada, and he didn't want to come back and do research. When his contract in Indonesia ended, he had an opportunity to do similar work in Madagascar. But he figured that if he stayed in shrimp aquaculture, he would begin to lose his connection to family back in Canada, so he and his wife decided to come home.

"My in-laws sent us a pamphlet about living in BC, and off the map, we picked Nanaimo, packed up our bags and showed up with ourselves and two dogs, and that was it," Ayranto remembers. "I looked around for some work and found work in salmon aquaculture. Salmon aquaculture is very unsupported in Canada, and there is a fair bit of controversy about it on the coast, so it was a pretty tough gig, and I got doing permitting (we were expanding on the coast). Engineering, First Nations—it was a real uphill battle, and there was a lot of foreign investment, and you had a real ceiling on where you were going to go. Unlike mining, there was not a lot of government support, financial support, for research and development. It was all coming from Europe—Scandinavia mostly. So I thought long term, this wasn't where I need to be."

Ayranto's wife's employer, Canfor, had just sold their Vancouver Island operations and had moved her to Vancouver while he remained on the island. He surveyed the job market and saw that mining had potential. Canada is a powerhouse with a lot of head offices, especially in the Vancouver area, and there were some real opportunities. He knew some people like Paul Gray in the mining field, so he started calling around. Through a couple of friends, he was connected with StrataGold, which at the time had the project in Guyana—it was going to be a mine—and they needed somebody to take care of the environmental assessment.

Ayranto went in for an interview. Terry Tucker was CEO at the time and Bob McKnight was the executive vice-president. He had breakfast with McKnight in Vancouver and in a couple of days had a job offer. He gave notice and moved to Vancouver on a Thursday. He and his wife bought a house on Friday and on Saturday morning he was on a flight to Guyana. It took a few weeks for him to figure out that there was no mine in Guyana; it was really a highly speculative venture. They had a couple of geologists on it, but in terms of advancing a project, they didn't have teams like John McConnell would later put together for Dublin Gulch.

Chad Williams, president and CEO of Victoria Gold from 2009 to 2011.

So Ayranto was drawn into managing technical work; he didn't know much about geology, but engineering and the other aspects were straightforward. With a scientific background, he could keep up enough to know if the company was on track or not. They were doing a prospectus financing, so they had to get all their technical documents up to speed and file them with the Securities Exchange: "We had our own guys' technical reports. The Securities Exchange turned down two of the three of them—for various reasons. I wouldn't say they were not technically complete, but they didn't conform to the prescribed nature in which you need to put a technical report together. And we had a short time, a few days, in which to put these reports together again.

"So myself and two other guys—one was Hugh Coyle—sat in the office for three days. We didn't leave. We slept there—a couple hours' sleep and we would be back up and rewrote these things. Three days later we got them finished, they all got signed off, and through that I got a pretty sizeable promotion. I guess I started out as VP environment or something like that, and ended up being VP of corporate development or some other title. And that's how I got my start in the business, really."

Top: StrataGold exploration drilling at the Eagle property, August 2005.

Bottom: The old exploration camp on the Eagle property in 2009.

A year and a half after Ayranto began working for StrataGold, the 2008 financial crisis hit. He'd helped StrataGold do all the right things with the Guyana property, but they couldn't raise any more money, so when Victoria Gold bought them out, it acquired StrataGold's Dublin Gulch property and also Ayranto, who had a management contract with StrataGold. Chad Williams refused to buy out the management contract. According to Ayranto, "I was working with the CEO half the time, and the other half I was fighting him on the contract. I really owe Chad a bottle of Scotch, because at the end of the day, what that facilitated was I was in the office for six or seven months or whatever, and I had a chance to spend time with John McConnell.

"John could see my skill set, and there were still lots of needs within the organization. It was a tough period, but John would go to lunch with me, or find other ways to spend time with me. Of course, it was a job interview—I kind of figured that out after a bit. The contract stuff wasn't working, so I had advice from a lawyer to give them notice, just walk away. So I said to John over lunch, 'This is coming to an end here pretty quick.' A couple days later, John said, 'You can come work with me. We'll make you an offer and keep you on the team.' That was in 2009 and he hasn't got rid of me yet."

As soon as Victoria Gold acquired the Dublin Gulch claims (which they renamed the Eagle Gold Project), they commenced further exploration and started the permitting process. The summer of 2010, they had two diamond drill rigs working on the Dublin Gulch property. Drilling has historically shut down in October due to weather conditions; however, in order to extend the drilling season and start the preparations of the site for eventual mine construction upon the receipt of permits, the company acquired and planned to install a two-hundred-person, all-season camp facility by the end of the year, for a total cost of approximately five million dollars.

But work at the property was not progressing well. Frustrated, McConnell put Ayranto on the project in August 2010. According to Ayranto, McConnell was alarmed

The core shack at the Eagle property, November 2011.

Top photos: Installing the one-hundred-person camp in December 2010.

Bottom: Aerial view of the one-hundred-person camp at the base of the hillside where the Eagle resource is located.

that the camp wasn't completed. "It was a 100-person ATCO camp—a five-or-six-million-dollar camp.[178] We were already into August; it would start hitting fall pretty quick, and John said, 'We still haven't got our pre-feasibility done. It's slowing down getting the environmental assessment application in.'" So Ayranto got the ball rolling. As he describes, first he phoned up his team and said, "'We need a camp here this year. We've got to get it installed; we've got to get it on site.' 'There's no way we can do it,' they said. 'We're running into winter and stuff.' And I said, 'Well, we're going to.' And we immediately got the earthworks going. We installed that camp in December 2010. There were terrible conditions—we were living in wall tents—but we got it done. We had some consequences—we had some shifting in the spring as you normally do—but I made sure that it got done."

The other major hurdle that had to be overcome, and in a hurry, was the environmental assessment. Ayranto phoned up the entire environmental assessment team and got them together. He asked them to put together the absolutely shortest schedule they could, with no contingency, no slack, no extra work, only what needed to be done. They spent a few days putting the schedule together and said that they could have it ready by December 20. But with absolutely no leeway—it would be extremely difficult.

So Ayranto said, "That's great work, guys," and phoned up McConnell on his cellphone with everyone still in the room. He said, "John, we're going to put this environmental application in on December 20." And he hung up. They finally submitted it on December 21, but only because the Air North plane they were using to ship the application couldn't fly on December 20 due to mechanical problems.[179]

Those were the two key challenges where Ayranto realized, "I'd better prove myself early," and he did. He also gives McConnell high praise for his management style: "He's really assured that unless I screw it up, I've got the project. There are others who are probably more qualified than I am, but John's loyal as long as you are pulling your weight. The other thing is he gives you lots of time and space to fix your problems. He's done it himself a few times, and he sits very tight in the saddle as you're heading towards a cliff. I know of no other individual who has that courage to stay the course: don't let the team get distracted—it's super important to stay focused. So it's been great. I've been working with John and Marty Rendall for ten years, and it's been one heck of a journey."

A report issued the following year revealed that the 2009 drilling program had boosted the potential value of the property to 154,294,000 tonnes, at a grade of 0.65 grams of gold per tonne.[180] More exploration was planned for 2010 with the aim of expanding the quantity of mineable ore in the Eagle zone, and by further defining the nearby Olive and Shamrock zones of mineralization. Victoria Gold had drilled twenty-eight holes totalling about 6,200 metres since the 2010 exploration campaign began in June. These holes were drilled mostly in the Eagle, Olive, Shamrock and Steiner zones. Due to the unusually heavy volume of exploration activity in the Yukon, assay turnaround was slower than expected.

Meanwhile, the environmental assessment process for the project was initiated, and talks with the First Nation of Na-Cho Nyäk Dun began to move forward. A lot of community liaison work for the project had been started by New Millennium/First Dynasty in the 1990s, but StrataGold hadn't really talked with the First Nation, so Mark Ayranto drove into Mayo one day and dropped in on the administrative office. At the first meeting, the

Stream characterization work for the environmental assessment, September 2009.

Community open house and dinner in Mayo for the Eagle Gold Project, December 2009.

Community consultation open house in Mayo in 2010.

As part of the community consultation process, Mayo community members visit the Eagle Gold Project site in 2010.

Nation was represented by Simon Mervyn, Anne Leckie and a few other councillors. They began to put an exploration agreement together, but it was not a straightforward path. Establishing a good relationship with the First Nation worked for Victoria Gold because they started early and they worked hard on it. Victoria Gold incorporated its importance into every level of the organization.

"It's truly not rocket science," declares Ayranto. "The best thing you can do is spend time in a coffee shop—there isn't one in Mayo, so whatever the equivalent is. People see that you are a regular person and you don't have horns in your head. The other big thing is when you go into a community, don't try to ram it down their throat. Pay attention and listen to what they have to say and figure out how you can contribute positively to that. Sometimes it works and sometimes it doesn't.

"I've been kicked out of several meetings—it's the exact opposite of a marriage. If you have a troubled marriage, it starts out happy and gets to troubled; well often with First Nations projects, it starts out troubled and you get to the happy bits. It's a lot of work but very rewarding."

Ayranto assembled a team to work together with the First Nation of Na-Cho Nyäk Dun that included Letha MacLachlan, who handled legal matters, and Sally Howson, who was the community liaison.

Letha MacLachlan, Q.C., Victoria Gold board member.

MacLachlan is a specialist in environmental impact assessment, permitting and the land side of things—land leases, land use permits, water permits and licences, and negotiation of agreements with affected Indigenous communities. She also has an unusual connection to Dublin Gulch. Her great-grandfather, Dugald MacLachlan, had been to Dublin Gulch in 1912.[181]

"I never knew him and my own father never really spoke of him. I only knew that he had travelled to the Yukon," explains MacLachlan. "I vaguely knew he was playing some role as a mining recorder. And then I started working for Victoria Gold in approximately 2009 when they were putting together their environmental assessment documentation. I went up to the mine site, came back to Whitehorse and sent an email to my cousin who is the genealogist in the family. I asked, 'Wasn't our great-grandfather up here some time ago? Do you have any information on that?' He had access to those diaries. He sent me a photograph of him standing beside the Olive adit, on which is written, *Olive Adit Dublin Gulch, Yukon*.[182]

"I was really gobsmacked when I found out that's where he had been, the day after I had just been there a hundred years later. I lived for decades in the Northwest Territories, and mining in particular has been a strong interest of mine. Even though I never met him, I often wonder whether or not my interest in the North had been handed down in some way. And now I'm probably as old as he was when he was here!"

Ayranto had to work hard to convince Sally Howson to join the team. She would be the first to tell you that she did not set out to have a career in the mining field. She describes herself as a "downtown" girl; she never talked to anyone about geology as a career when she was growing up. She didn't really need the Dublin Gulch work, but she liked the project because she had a connection to it from years before.

Her studies in university were in urban geography, but after she graduated, she was adamant that she did not want to become a teacher. She decided she wanted to work in the Yukon. Through friends of her boyfriend, she learned that she could get a job working on a project in the south Canol region of the Yukon, and signed on for a summer as a "dirt-bagger." A dirt-bagger collects soil samples to be shipped to a laboratory for analysis. The pay was good and the work was seasonal, which suited her just fine, and she continued to get jobs in the north.

Because she had a degree in geography, the people who hired her assumed that she knew more about geology than she actually did. She had other qualities that kept her employed: she was a hard worker, she liked camp life, she got along with people and she learned quickly. These attributes served her well, and she found steady employment, even during the downturns in the economy when more highly trained and skilled professionals were out of work. She and her husband worked on projects in northern Canada and on distant continents.

Then Howson went back to school and obtained a degree in environmental science, which got her a job working for Westmin at the Premier mine in Stewart, British Columbia. But she didn't want permanent employment at the Premier Mine, so she only filled in for a few weeks. Then Hans Smit, the man who hired her, said there was an opportunity to work on the Dublin Gulch project, when the property was still owned by New Millennium, doing the environmental study and community work.

There was a big *but*: the work wouldn't be full-time, so she would also have to work on the drill crew that was doing exploration work on the property. So she decided to work on the drill crew at night and do the environmental work during the day ("You don't count the hours or you start driving yourself crazy"), hoping that she would transition to the environmental work full-time when they started the permitting process.

Howson worked with Hans Smit and Steve Stein, the vice-president of operations for New Millennium/First Dynasty. Stein was responsible for completing the feasibility and engineering studies, and the environmental assessment, for Dublin Gulch during the 1990s. They also needed to get the First Nation and the community of Mayo on board. "When I worked with New Millennium/First Dynasty, we started negotiating what was called an Impacts and Benefits Agreement," recalls Howson. "Then everything stopped in 1997 because of the gold price. We were starting the negotiations with the First Nation of Na-Cho Nyäk Dun at that point. They had appointed two people for their committee, and then Steve Stein and Hans Smit and I were there for the company. So we had started negotiations."

Victoria Gold representatives and First Nation of Na-Cho Nyäk Dun citizens visit the Fort Knox gold mine in Alaska in 2009.

The environment was always the big issue for the First Nation of Na-Cho Nyäk Dun. The people from Mayo had the United Keno Hill mine in their backyard, and it had been operating since before environmental concerns were addressed. On the social side, they wanted to make sure that the community got as much benefit from the project as possible, because at the United Keno Hill mine, people in the community didn't necessarily get the jobs.

People were concerned about what a new mine might do to the water. They were very worried about cyanide, so New Millennium had to educate community members about what cyanide heap leach was. Fortunately, Brewery Creek mine was being built at that time, and the company was able to take council members and elders to Dawson to show them how the protective lining was put down.

Later, when Sally worked for Victoria Gold, they similarly took a group of government people and First Nation of Na-Cho Nyäk Dun citizens over to the Fort Knox mine in Alaska. Howson had done much in the community in the 1990s educating people about the heap leach process, which was a great benefit when Victoria Gold hired her to do community liaison work. It was a big advantage that people already knew her. "You know, people have long memories," Howson reasons. "I spent a lot of time in the community, a lot of time… Chief Robert Hager said to me, 'Didn't you use to live here?' I spent that much time up there.

"I think that is what people don't get about consultation and community engagement. It's not just going up and having a meeting every two months. I was very involved with the community, so I built a lot of bridges. I think when I came back with Victoria and I was part of the negotiating team, I felt that the community should benefit from it. It's a good place for a mine—it's already been placer mined, and it's far from pristine so if you are going to have a mine somewhere, that's a good place for it. Mayo is very aware of what mining can do for the community."

Even the diehard opponents eventually came around. Mayo resident Frank Patterson was one of them, and was one of the people to visit the Fort Knox mine. He came to understand that the company was sincere about making sure that the environment was protected, and that the community got the benefits. By December 2010, the company had submitted the final project proposal to the Yukon Environmental and Socio-Economic Assessment Board, with a completion of the assessment within two years. This was a significant milestone that engaged the formal Yukon environmental assessment review process, and was an important precursor to the permitting of full-scale mine development.

Victoria Gold board of directors at the project site, June 2010. From left to right: John McConnell, Len Krol, Hugh Agro, Chad Williams, Sean Harvey, Mike McInnis.

A pre-feasibility study for the company detailed the planned activities at Dublin Gulch for the development of an open-pit heap leach gold mine that they were now calling the Eagle Gold Project. Gold production would be approximately 170,000 ounces per year starting in 2013. The cost of production would amount to five hundred dollars per ounce of gold produced. "Victoria's submission of the project proposal is a major step toward the development of a gold mine at Eagle," said Chad Williams. "We would like to thank [engineering consultant] Stantec and the entire Eagle team for their tremendous work and effort in delivering the project proposal."[183]

Chief Simon Mervyn and John McConnell signed a memorandum of understanding on May 20, 2010, in Dawson City. "The First Nations people have used this area for millennia for traditional harvesting," said Chief Mervyn. "However, we recognize that there are economic benefits in modern times that we can choose to participate in."[184] The agreement expressed corporate resolve to maintain good communications with the First Nation, and to work with them in a progressive and positive manner.

Victoria Gold now had a one-hundred-person camp within a few kilometres of where the open-pit mine on the Eagle deposit was to be created. Portable generators supplied electrical power to the site in the short term while Victoria Gold negotiated a deal with the Yukon Energy Corporation. A power line paralleled the highway between Mayo and Keno City, and Victoria Gold had a letter of intent in place with the Yukon Energy Corp to supply grid power via a spur line to be constructed along the existing access road to the site of the proposed mine.

In early February 2011, Chad Williams, who had been president and CEO of Victoria Gold for three and a half years, stepped down from his position. During his tenure, he had engineered the acquisition of StrataGold Corporation, advanced exploration on their Nevada properties and saw the delivery of a pre-feasibility study for the Eagle Gold Project. John McConnell was appointed as Williams's replacement.

John McConnell was born into a mining family in Nelson, British Columbia, and grew up in a mining town. His father was a mine electrician at the Emerald mine near Salmo and Trail in the southeastern corner of British Columbia. "I started working there, and after high school, worked in the mine as an

John McConnell became president and CEO of Victoria Gold in 2011.

underground miner for a number of years," McConnell recalls. "Finally, my father convinced me I should pursue a higher education, so I went to the British Columbia Institute of Technology [BCIT] and became a mine technologist." McConnell went directly from BCIT to the Colorado School of Mines in Golden, Colorado. It was a good choice. Both a teaching and research institution, the Colorado School of Mines was established in 1873 and is often ranked as the best mining school in the world.

After graduating with his degree in mine engineering, in 1980 he found employment with Strathcona Mineral Services at the Nanisivik mine on the northern tip of Baffin Island. McConnell earned his spurs as a mining engineer at Nanasivik: "The first couple of years I was in Nanisivik, you were referred to as an 'engineer in training.' I spent time as a ventilation engineer, and then as a mine planning engineer, and spent time actually working underground with the equipment, so it was a good apprenticeship.

"I guess I've always had the ambition to work my way up the corporate ladder; I think people would say that I'm extremely hard-working. I think back to the days of Strathcona Mineral Services and later when I was consulting, and you had to get a report out by a deadline. I'd work forty-eight hours straight to get it done. I would sleep on my desk, that sort of thing. My wife Tara would tell you I'm still sort of like that. It requires long days and long hours, and I put the time in."

One of the first things McConnell did after he was promoted to CEO was move to Whitehorse. There are several strategic advantages to living within the jurisdiction you are investing in. Before moving to Whitehorse, he had only met the Yukon's premier once. But living in Whitehorse meant he would bump into government leaders and key officials in the grocery store or at the movies or on the golf course. "It just made that relationship with permitters and regulators much different," McConnell says. "I wanted to be part of the community and for Victoria to be seen as part of the community."[185]

Tara Christie, Yukon mining executive and John McConnell's wife.

But for McConnell there was an additional incentive. A year earlier he had met Tara Christie, who had grown up in a placer mining family that operated outside of Dawson City on Scroggie, Mariposa and Black Hills creeks. Her family moved to the Sixtymile district before finally settling in Indian River. As a teenager, she operated an excavator, cooked and did "all the things a placer mining child does." She went to the University of British Columbia in 1992 to become a geotechnical engineer and left school six years later with a master's degree.

"I wrote my thesis on the night shift, running a half-shift and running a loader—which I hated. I wrote it in a little trailer down by the plant. I did not think I was going to get my thesis done," she remembers. But she did, and began to do consulting work. She also became the president of the Klondike Placer Miners' Association during some troubled times for the industry. She was drawn into and became very active in the BC/Yukon Chamber of Mines and the Prospectors and Developers Association of Canada.

McConnell first saw her giving a presentation at a mines ministers' conference in 2003 or 2004. But they didn't meet until seven years later, at a one-day conference in Dawson City sponsored by the Yukon Geological Survey. He walked over and introduced himself to her at Diamond Tooth Gerties Gambling Hall, and they went out on their first date a week later. There were some down years that followed, with less activity at Victoria Gold, so McConnell came to Dawson City to work on the Christie family placer mine and be closer to Tara. It was there that he learned how to pour gold bars and improve mine efficiency.

Today, McConnell and Christie are the ultimate power couple in Yukon mining. People sometimes get the impression that she is actually on the Victoria Gold payroll, but she isn't. Christie was on the Yukon Environmental and Socio-Economic Assessment Board when they were dating, so it posed a potential conflict of interest for her. Fortunately she wasn't on the executive committee, and removed herself from all discussions regarding Victoria Gold.

According to Christie, it has helped the relationship to both understand what the busy mining lifestyle is all about: "It's nice to share the passion and understand what's going on in your life, and why you are stressed out and busy, and what the implications actually are. It would be hard to be in the business with his travel schedule and everything that's going on if I wasn't sympathetic, if I didn't understand what is going on. You can see how hard this lifestyle is on relationships because of the drama of it all and the need to travel. And the demands, and the times that you have to work unbelievable hours. Family vacations turn into work. I am no stranger to this and that's why we get along."

2011 became a pivotal year in the development of the Eagle Gold Project. On January 25, Victoria Gold announced that Wardrop, a Tetra Tech company, had been hired to complete an NI 43-101-compliant feasibility study for the Eagle Gold Project by the fourth quarter of the year, for a cost of four million dollars. Wardrop was an internationally recognized engineering firm with considerable northern mining project experience. The year before, Victoria Gold had announced the results of a preliminary feasibility study that highlighted the development of a mine at Dublin Gulch, with production to start in 2013 of an estimated 170,000 ounces of gold per year and cash operating costs below US$500 per ounce.

Wardrop incorporated previous work, including updated resource and reserve estimation, principal mine design, planning and financial modelling and engineering. Wardrop led the development team of BGC Engineering; Kappes, Cassiday & Associates; and Stantec, each of which was involved in the pre-feasibility study. In early February, McConnell announced that Victoria Gold and the Yukon Energy Corporation had signed a letter of intent setting out the details of a Power Purchase Agreement. Under this agreement, Victoria Gold aimed for a secure commercial rate for the life of the mine. Ensuring a reliable source of power at a reasonable rate would be one more element in making the mine a viable proposition well into the future.[186]

Safe haven for winter exploration, October 2011.

Above: The 2011 engineering program to locate a groundwater source to supply water for the heap leach facility.

Facing page: The 2011 exploration program.

A helicopter moving drill supplies for the 2011 Rex-Peso exploration program.

Thanks to Victoria Gold's strong cash position, Lyncorp Drilling Services of Calgary was on-site by the end of February to initiate a nine-million-dollar drilling exploration program at Dublin Gulch—three months earlier than usual, thanks to the newly installed all-season camp. The results of the 2010 drilling season produced an optimistic increase in indicated and inferred reserves that could only stimulate interest in the potential for Dublin Gulch. The 2011 exploration program, which would be Victoria Gold's most ambitious to date, focused on expanding the known limits of the Eagle gold deposit. In addition, the program included extensive exploration of the neighbouring Potato Hills Trend, as well as the Olive and Shamrock targets, which had shown promising drill results in 2010. Finally, the 2011 exploration sought new and promising targets across the extensive and largely unexplored block of claims, adding more than seventeen kilometres of cores from more than one hundred diamond drill core and reverse-circulation drill holes.[187]

In June, Victoria Gold and the First Nation of Na-Cho Nyäk Dun signed an extension to their exploration agreement, which was superseded later in the year by the signing

First Nation of Na-Cho Nyäk Dun chief Simon Mervyn (left), FNNND council members and Victoria Gold chief operating officer Mark Ayranto at the Comprehensive Cooperation Benefits Agreement (CBA) signing ceremony, October 17, 2011.

of a Comprehensive Cooperative and Benefits Agreement (CBA). Chief Simon Mervyn commented that "The completion of this agreement... has advanced the development of a major gold mine while promoting the socio-economic and environmental objectives of the Na-Cho Nyäk Dun."[188] While providing certainty for developing the Eagle Gold Project and expanding the exploration program, the agreement provided the First Nation of Na-Cho Nyäk Dun with employment and economic development opportunities while respecting and promoting their desired environmental protection objectives. Among other things, the First Nation received further employment and training opportunities, and the agreement created a protocol for ongoing communications between the two parties.[189]

Finally, just before Christmas, Victoria Gold and the First Nation of Na-Cho Nyäk Dun announced the signing of a letter of intent for the staking of Na-Cho Nyäk Dun land adjacent to the Dublin Gulch project, nearly doubling the land package within Na-Cho Nyäk Dun traditional territory in the Yukon. Victoria Gold staked additional claims on two parcels of Na-Cho Nyäk Dun category "B" land adjacent to, and to the west (1,012 claims, Victoria Block West or "VBW") and south (473 claims, Victoria Block South or "VBS") of, the existing Dublin Gulch property. This ground, which covers an area of 290 square kilometres, had not seen any exploration activity in over twenty years. This agreement boosted Victoria Gold holdings to 3,408 claims covering 646 square kilometres.

The environmental assessment process was also progressing. In July, Victoria Gold announced it had cleared the first hurdle when the Yukon Environmental and Socio-Economic Assessment Board deemed the project proposal for the mine to be adequate. This meant they could move onto the screening phase. If successful at the screening stage, Victoria Gold could apply for its water use and quartz mining licences. The feasibility study was nearing completion and was expected to be released early in 2012.

Meanwhile, Kelly Arychuk was appointed the company's vice-president, mine support services. With years of experience in the human resources field and knowledge of the mining industry, she had the skills needed to build the larger team necessary for construction of a mine. During the year, the Victoria Gold team had expanded from ten employees to thirty with the addition of two new members to the board of directors: Edward Dowling from Alacer Gold Corporation and Christopher Hill from Aecon, Canada's largest infrastructure development company. Both brought additional experience and knowledge to the board, increasing the expertise necessary to advance the development of the Eagle Gold Project.

To support this aggressive development of the Eagle Project, Victoria Gold raised thirty million dollars by issuing a combination of common shares and flow-through

shares to finance further exploration and corporate activities. John McConnell saw this as the bridge financing that was needed to carry the project through to a final project financing decision and the start of construction.

The long-awaited feasibility study by Wardrop which had been initiated the previous year, was finally released in February of 2012. Buoyed by astronomical gold prices, the results of the study were positive. In June of 2010, gold was standing at $1,100 per ounce. Within five months, it was approaching $1,400 and things were going to get much better. A month later, it peaked at over $1,600. By August 2011, it had broken through the $1,800 mark and in the following months it went on a frenzied roller-coaster ride, cresting at over $1,600 several more times before September of 2012.

The study projected an open-pit mine that would deliver almost thirty thousand tonnes of ore to a three-stage crushing mill each day. The crushed ore would be stacked on an in-valley heap leach pad, and the gold recovered from an adsorption/desorption gold recovery (ADR) plant. Average annual production over a ten-year lifespan would be nearly two hundred thousand ounces per year. The capital cost for construction would be $382 million, and the operating cost would run to about $542 per ounce. The sustaining capital cost would add another $132 million and closure costs would amount to $64.2 million.

The start-up date when construction was complete was forecasted to be the fourth quarter of 2014. The payback on the initial capital cost outlay would be a little over three years. An all-weather road to the site, a one-hundred-person camp already installed and a letter of intent with Yukon Energy Corporation all added to the attractiveness of the project.[190] Meanwhile, there were two drills on-site at Dublin Gulch further testing the near-surface high-grade targets and further defining the extent of the ore body. Within the year, the company anticipated the completion of the environmental screening and the granting of a quartz mining licence, which would allow construction to proceed.

As development progressed, Victoria Gold sharpened its focus on the Dublin Gulch property by selling off some of its Nevada assets. The high price of gold made these properties even more attractive. The first to go was the Relief Canyon property in Pershing County, Nevada, in early April. The transaction would transfer Victoria Gold's interest in this Newmont Mining Corporation property to Pershing Gold Corporation for two million dollars cash and four million dollars in Pershing common stock.[191]

Facing page: Auger drilling at Eagle on the Dublin Gulch property to support geotechnical work and ground condition characterization for future facilities, September 2011.

This transaction was followed quickly by another, the sale of Victoria Gold's Cove McCoy property to Premier Gold Mines Ltd. The deal included an immediate cash payment of six million dollars upon signing, followed by payments totalling an additional twenty million dollars over two years and further payments of up to twenty million dollars should the property go into production.[192] The next to be sold off was the Mill Canyon property to Barrick Gold for fifteen million dollars in cash and another nine million dollars in contingent payments and property exchange. In total, these three transactions infused another forty-nine million dollars in cash into the Eagle mine venture, with an additional twenty-five million dollars million in contingent payments plus royalties.[193]

Victoria Gold now had working capital on hand totalling $56.7 million, which provided ample cash to advance the permitting and engineering studies for the Dublin Gulch property. And more good news followed. In early September, the Yukon Environmental and Socio-Economic Assessment Board completed a draft screening report for the Eagle Gold Project. "As a result of this assessment," the report concluded, "the Executive Committee recommends to the Decision Bodies that the Eagle Gold Project be allowed to proceed without a review, subject to terms and conditions identified in this Report." The environmental assessment would be completed in early 2013, after which permitting for the mine could commence.[194]

Victoria Gold had been negotiating an access and exploration agreement with the First Nation of Na-Cho Nyäk Dun since December of the previous year. In September, they were able to draw this to a conclusion by signing an agreement that brought certainty to the further exploration activities of the company, while establishing an avenue of communication between the two parties. Victoria Gold would provide financial support to the First Nation while ensuring that the First Nation of Na-Cho Nyäk Dun would have employment and economic opportunities. Victoria Gold would also respect the environmental objectives of the First Nation.

The signing of this agreement was the first time First Nation land would be explored in the Yukon with the prior consent of the Nation. Chief Simon Mervyn applauded Victoria Gold for their commitment to working with the First Nation of Na-Cho Nyäk Dun and for respecting the land and the environmental values put forward within their traditional territory.[195]

2013 promised to be an eventful year, but it also proved to be one with challenges. John McConnell was optimistic about the early results of exploration work in the newly staked blocks of property to the west and south of the existing block of claims covering Dublin Gulch. "These results demonstrate the very large scale of the wider Dublin Gulch property," he stated, "and provide promising indications of gold mineralization adjacent to the company's flagship Eagle Gold deposit."[196] A pleasant surprise for the project was contained in the final 2012 drill results on the property, which suggested that the Olive zone, only 2.5 kilometres away from the Eagle Gold Project, also had potential for heap leach.[197]

Victoria Gold sold another Nevada property, Big Springs located in Elko County, to MRG Copper LLC for the purchase price of up to six million dollars US, which consisted of four million dollars in cash as well as another two million in contingent payments. Later in the year, Victoria Gold would receive a second payment of ten million dollars from the sale of the Cove McCoy property the year before.

Further big news came in February when the Yukon Environmental and Socio-Economic Assessment Board issued its final screening report, recommending that the governments of Canada and the Yukon allow the Eagle Gold Project to proceed. The decision, which was more than three hundred pages long, laid out a list of 123 terms and

Mark Ayranto (third from left) and Yukon Water Board and Secretariat members tour the Dublin Gulch site in October 2013.

conditions that had to be met for Victoria Gold to be in compliance. Many of these dealt with the safe handling and use of cyanide, as well as water quality and protection of permafrost, fish, wildlife and habitat.

On April 9, 2013, the Yukon government issued a press release announcing the filing of the decision document for their approval. A week later, the federal government similarly filed their decision document approving the project to move forward. Five months later, the quartz mining licence was issued to Victoria Gold, bringing three years of environmental assessment to a conclusion.[198] "Eagle is now one of the few major gold projects in the world in a first-class jurisdiction that is shovel-ready, pending financing," said McConnell. [199]

M3 Engineering was chosen to complete the detailed design work for the Eagle Gold Project. M3 had relevant experience in the Yukon, having recently completed a feasibility study for another large-scale Yukon mine project. With detailed engineering for the project, thirty-five million dollars in the bank and a growing team to implement the work, Victoria Gold was ready to move forward. The application to amend the existing water use licence so that construction could proceed was to be filed before the end of 2013, with receipt of the amended licence expected in 2014. In addition, Victoria Gold was planning to apply for a new water use licence to support operations at the Eagle Gold Project, which were expected to begin sometime in 2015.

But not all was roses and sunshine for the company. After reaching historical highs, the price of gold began to tumble through 2013. Capital markets became sluggish and the share price of Victoria Gold stock remained severely undervalued despite the state of readiness of the property for development. The company responded over the following two years by reducing its core staff from thirty-five employees to ten. "We are disappointed equity markets are unsupportive to begin construction this season," explained John McConnell. "However, Eagle is a rare shovel-ready project and the ore body does not go away. We will continue to de-risk and enhance the overall project to facilitate a quick and efficient site mobilization in 2014."[200]

Work proposed in the interim included road and bridge improvements. The Yukon government committed one hundred thousand dollars to improve grades and drainage. It also built a new bridge across the Haldane River, which brought the river crossing up to highway standards. Further advancement of the project included detailed engineering studies to meet permit requirements, and detailed engineering to support the start of construction in 2014.

Top: Yukon government upgrades to the Haldane Bridge in 2013.
Bottom: Local Mayo contractor Ewing Transport working on access road improvements.

Construction, however, did not begin in 2014. Instead, attention was directed to another promising zone of mineralization within spitting distance of the Eagle proposal: Olive. Noteworthy here were the shafts and adits from prospecting of the deposit a hundred years earlier. Previous trenching and drilling from 1991, 2010 and 2012 had already provided promising results. The goal was to define a mineable ore body within this zone that would be suitable for the heap leach technology. Phase one of this exploration program, which included three kilometres of diamond drilling, began in May. The results of this work were encouraging, suggesting that the ore was of a higher grade than that in the Eagle zone. Phase two of the 2014 exploration program began at the end of July.

The phase two program involved further delineation of the high-grade Olive zone and more exploration drilling to the northeast and southwest of the phase one drilling. The 2014 exploration program included forty-nine exploration diamond drill holes, twelve metallurgical diamond drill test holes, seven geotechnical drill holes and 882 metres of surface trenching resulting in 6,757 new assays.

By November the results were in. Victoria Gold was confident there was an additional ore body to be mined, and planned to develop an NI 43-101 resource estimate for the Olive zone. Metallurgical testing by Kappes, Cassiday & Associates of Reno, Nevada, revealed that the Olive zone, though smaller than Eagle, was mineable because of the higher grade of the ore. This added to the overall potential of the proposed mine and opened the door to further exploration along the Potato Hills trend that might extend the life of the mine.[201]

A final piece of good news came at the end of 2015, when Victoria Gold received its water use licence allowing for the construction, operation and closure of the Eagle Gold Project. During the week-long public hearing in June 2015, the water board chair stated, "Out of many of the companies or applicants that come before this board, I think you [Victoria Gold] are remarkable for having had with you—for the duration so far—virtually the same team. You've put together an 'A' team and that in turn has led to your preparation of what in our opinion is a very, very good submission. And that goes a long way to assisting the entire application process. It also goes a long way to giving evidence of the trust that exists between yourselves—as the applicant—and the people in this community, and larger, the people in the Yukon Territory. So, congratulations. You've done a remarkably good job in bringing your team together and providing a good application. That's going to stand you well as you move forward."[202]

Top: Water use licence hearing with the Yukon Water Board in June 2015.
Bottom: Attendees included (left to right) Mark Ayranto, Warren Newcomen,
Marty Rendall, Jim Harrington, Timo Kirchner, Troy Meyer, Scott Tinis, John McConnell and Sally Howson.

2016 began with McConnell announcing the exploration program for the year. Using $3.6 million in flow-through financing, Victoria Gold planned to focus on the Olive and adjacent Shamrock mineralization zones, and test the intermediate area. This new data would inform the preparation of a new NI 43-101 resource estimate, which would include, for the first time, the Olive ore body. The drilling contract for the 2016 Olive-Shamrock exploration program was awarded to Kluane Drilling Ltd., a family-owned and -operated Yukon-based company with global operations.

Drilling commenced in March. Three diamond drills were up and running and forty personnel were on-site in the company's all-season construction camp within days. Olive and Shamrock were shallow deposits, so could be mined using existing infrastructure. The program consisted of diamond drilling, surface trenching and geophysical surveys over the Olive-Shamrock zone, with a focus on the previously undrilled areas linking Olive and Shamrock mineralization. The exploration program would concentrate on expanding the length of confirmed near-surface, high-grade gold mineralization within the Olive-Shamrock shear zone trend and target the previously untested three-hundred-metre gap between the Olive and Shamrock mineralization.

The results from the drill program started coming in within weeks of commencement. All were encouraging, including the results from the drill holes in the area between the Olive and Shamrock zones, which demonstrated that they were in fact a single ore body a kilometre in length. The results of this testing would be integrated into resource estimates scheduled for release in the fall of 2016.[203]

In April, Victoria Gold announced that JDS Energy and Mining Inc. had been engaged to lead the feasibility study update for the Eagle Gold Project. With offices in Canada, the US and Mexico, JDS is a company that has undertaken major projects around the world. The update was designed to account for several positive changes.

In the previous eighteen months, Victoria Gold had received a water use licence, hence the Eagle Gold Project was fully permitted to begin construction. A weaker Canadian dollar, lower fuel and equipment costs and an improved labour environment all enhanced the viability of the project. The feasibility study was released in October, including the Olive zone for the first time. It boosted the estimated gold reserve of the property to 2,663,000 ounces. The rate of production was now projected at 33,700 tons per day, with a strip ratio of 0.95:1.[204] The capital cost for construction was $369 million.[205]

Victoria Gold also seized the opportunity to purchase a used all-season camp complete with 110 dorm rooms, industrial kitchen, recreational and mud rooms, and arctic

The expanded Eagle camp.

An airborne geophysical survey over the Shamrock zone.

corridors for the price of $275,000. This transaction amounted to a savings of nearly six million dollars compared to the purchase of a brand-new camp. Located less than one hundred kilometres from Dublin Gulch, the camp was in excellent condition and required only minor refurbishments. And the cost and logistics of moving the camp to the site were relatively modest, adding to the one-hundred-person all-season camp already there.

Victoria Gold not only had to save cash; it had to generate it. In April, Victoria Gold raised twenty-four million dollars in a private placement, selling eighteen million dollars' worth of shares to Electrum Strategic Opportunities Fund, a Madison Avenue firm, and six million dollars in shares to Sun Valley Gold, a privately owned hedge fund sponsor from Ketchum, Idaho. As a result of the transaction, Electrum held 13.6 percent of all the common shares in Victoria Gold. In June, Electrum exercised its right to nominate one person to the company's board of directors, and professional engineer Heather White joined the board. Meanwhile, Sun Valley, which had already sunk a considerable amount of money into Victoria Gold, held 18 percent of all Victoria Gold stock.

During the summer, Victoria was successful in raising another $28 million through a share issue that was underwritten by a syndicate of banks.[206] By the latter half of November, the company had raised $4.7 million by way of a flow-through share offering, which brought the financing for the 2017 exploration season to more than $6 million. With this funding, Victoria Gold planned to continue testing the Olive-Shamrock zone, while expanding the program to other promising features that had been named East Potato Hills, Steiner, Nugget, Rex-Peso, Lynx Dome and Falcon, all of which were within five kilometres of the Eagle zone. The prospects for the forthcoming year were looking good.[207]

2017 was a busy year for the Eagle Gold Project as Victoria Gold moved closer to the major construction phase of the mine. The work that lay ahead was long, arduous and expensive. Massive earthworks for water containment and the heap leach pad had to be built. Heavy equipment would have to be ordered long in advance of arrival on-site. The ground would have to be prepared for the on-site road, crusher foundations, gold processing facility and conveyor system, which would be built as the project progressed. Contracts would have to be signed for project management, earthworks, surveying, road construction and various other services. Arrangements would have to be made for a hydro line to supply the mine with energy.

This work was going to take a lot of capital, so McConnell went looking for financing: "Once we completed the feasibility study in 2016, we approached a number of banks that generally lend money to mine development. None of them are Canadian—there are two Australian banks, the Commonwealth Bank and the Macquarie Bank, a couple of French banks, Société Générale and BNP Paribas. We approached them and got a reasonably warm reception. All of them have been to the site, and we chose BNP Paribas."

Meanwhile, the $6.3 million exploration program was moving ahead to test a number of targets, including East Potato Hills, Rex-Peso, Nugget, Lynx Dome and Falcon (Falcon is located in the block of claims, designated VBW, to the west of the Dublin Gulch claim block.) None of the development could proceed until the financing of the project was in place. In January, Victoria Gold appointed BNP Paribas as the sole mandated lead arranger to raise US$220 million of senior, secured project debt for the Eagle Gold Project. Further details had to be ironed out, but the closing of the deal would take place by the end of June.[208]

A few weeks later, Victoria Gold appointed the man who would make the mine a reality, the vice-president of project execution, Tony George. George's great-grandfather was a Cornish tin miner who moved to South Africa with his wife in the 1890s. They came back to England later and farmed after that. So mining was in his blood, but a couple of generations back. His hometown is Camborne, where the world-famous Camborne School of Mines is located, but he points out, "If you ever grew up in Camborne, you'd know why I left. I went to the Royal School of Mines in London."

George graduated in 1981 and went straight down to South Africa to work for De Beers for five years before he moved on. De Beers offered an excellent graduate training program. He was fortunate to work in the underground mines—working on block cave sub-level caving—and in the open pits. He was a shift boss and mine captain for development underground at the Finsch mine, which is about a two-and-a-half-hour drive out of Kimberley, South Africa. It was a large open-pit operation that eventually became an underground mine. George was there during the stage where they were developing it underground.

Tony George, vice-president of project execution.

To advance in the mining field, you must move around and get a wide range of experience, and that is what Tony George did. This mobility was good for a while when his children were young, but when they became older, they needed some stability. A friend said, "Come to Canada. There's lots of jobs for mining engineers." He came to Canada in early 1993 and went job hunting. Within six weeks he had a job offer to be a planning engineer with the Iron Ore Company (IOC) in Labrador. When he got there, they said he would be the drilling and blasting engineer, which is quite a responsible position.

Moving from Africa, where the temperature was plus forty, to Labrador, where it was minus forty, was a bit of a shock for George and his family. He worked for IOC for two and a half years, achieving the position of superintendent of technical services before he left. From Labrador, he moved to Vancouver to work for Rescan for three years before he moved over to AMEC for five years and worked on about forty projects in twenty countries around the world. His last job with them was the pre-feasibility study for the Victor Diamond project in northern Ontario, just north of Timmins. After that, De Beers re-employed George to join the team through consulting and then through operations. He was the general manager through construction, but left before it was completed. Again, it was good experience.

George worked for Aura Minerals, Inc., for five years on various projects around the world, after which he spent seven years with the Lundin Group. They built the Karowe diamond mine in Botswana. They spent about $150 million on a mine that generates about $250 million of revenue a year. His last job with Lundin was to work on the feasibility study for the Fruta del Norte mine in Ecuador. He was vice-president of project development, a title that covered everything from feasibility studies through to construction of an underground gold mine. Fruta del Norte was a large, very complex project that George worked on for nineteen months. He took some time off after that project was completed.

"I had a couple of calls from John McConnell," he relates. "He said, 'Look, I think we can get our financing in place. Would you be interested to join the team?'" This was in November of 2016. McConnell had done a little background research on George, who was recommended to him by Patrick Downey, one of the board members at Victoria Gold. According to George, "After Christmas, no word from John. By that stage, I was looking for work, and had applied for a couple of things, and then John called in the middle of February, and said, 'Oh, come in for a cup of coffee, I want to catch up.'

"The first time, Mark Ayranto was there as well for lunch, for an hour. And the second time in February, for coffee, John says, 'Here is your offer letter.' I started two weeks later. They are a good group. John, Marty Rendall and Mark are the stalwarts of this project, and they have lived with it for seven to ten years, Mark especially. It's really their baby. I make sure that I come in on the sideline and do my little bit, but it's their glory. They really have put so much into it."

On March 27, 2017, Victoria Gold had awarded the engineering, procurement and construction management contract for the Eagle Gold Project to JDS in partnership with Hatch Ltd., a Canadian consulting and engineering firm. JDS-Hatch was to continue and complete the engineering work that had already begun after JDS's feasibility update was released in 2016. JDS-Hatch would then move into detailed design. The JDS-Hatch team had considerable experience working together on northern projects, including another Yukon project, Minto mine, and projects in Nunavut and the Northwest Territories.[209]

A day later, the company made another important announcement. It had entered into an exclusive agreement with Finning Canada, a division of Finning International Inc., to

Victoria Gold purchased a fleet of eleven Caterpillar 785D off-highway trucks, each capable of hauling 150 tons of ore, seen here assembled on-site at the Eagle Mine.

supply the mining fleet for its Eagle Gold Project at a price of roughly fifty million dollars. This fleet would include two massive 6040 FS hydraulic shovels, which were subsequently nicknamed "Beauty and the Beast;" eleven 150-ton, 785D off-highway trucks; and various auxiliary Caterpillar mining equipment. There is a decided advantage to purchasing brand-new equipment at the beginning of a mine operation, as the risk of mechanical breakdowns is greatly reduced. Victoria Gold and Finning further agreed to identify opportunities for used auxiliary Caterpillar equipment, as well as providing delivery date guarantees.[210]

Anticipating a larger on-site workforce as the mine moved into the construction phase, in April Victoria Gold set up the camp units that had been purchased the year before, doubling the camp capacity by increasing the number of rooms to 250 and increasing the size of the dining facility.

The company also announced that it had appointed a new vice-president of exploration, Paul Gray. Gray is a second-generation geologist—rocks seem to be in his blood. Paul Gray was born in 1974, an only child, and attended an all-boys private prep school in Pomfret, Connecticut. In his unrestrained manner, Gray describes it as "a school for rich kids to meet other rich kids to have rich babies." He wasn't part of this culture; he attended because it was nearby. Gray was surrounded by rocks, and the lure of rocks, from day one. He was also surrounded by computers. His father was a mathematical geologist who bought his first personal computer in 1980. From that point on, there was always a computer in the house, doing complex calculations.[211]

Gray originally enrolled at Dalhousie University in business, but after his first year he took the path of least resistance and switched to geology. It was at Dalhousie that he first met Mark Ayranto, since they both stayed in the same dorm. After obtaining his honours degree in geology, Gray sent out resumés to Vancouver and then hitchhiked there himself, where he found a geological job on his third day in. He secured a job with Alan Savage, whom he considers to be his mentor. Savage had just sold Canamin Resources after the discovery of the Huckleberry copper mine.

When Gray met him, Savage was ensconced on Howe Street. He had parlayed his money from the Huckleberry project into several companies. To avoid taxes, he started a company, and Gray was hired on the spot as his only employee to stake claims all summer just outside Kelowna. He spent three months staking them, six months working them, and the winter writing reports. He repeated that for a couple of seasons; meanwhile, Savage showed Paul how his entire corporate structure worked.

Paul Gray, vice-president of exploration for Victoria Gold.

Unlike his father, Gray spent the first ten years of his professional career doing field-work in the summer and working on Howe Street in the winter. Savage acquired some BC assets that had been sold off by Falconbridge at a low point in the market, and for a number of years, Gray "kept them alive" for him. He did the annual assessment work and learned all the tricks to work the system and keep the claims in good standing. When British Columbia went from manual claim staking to digital staking, Gray was appointed as the youngest member of the advisory committee.

He ended up in Honduras for a couple of years, then went looking for uranium in the southwest for about five years, worked in Mongolia for a while, and then worked in South and Central America. In about 2008 he started a consulting company, finding work here and there, mainly by word of mouth. Meanwhile he crossed paths once again with Mark Ayranto, who was in Campbell River growing shrimp. When Ayranto got a job with StrataGold, he hired Gray in 2010 to do some consulting work.

When Gray went to the Eagle deposit, the first thing he did was review all of the reports of previous work done there. He pored over the literature and tried to look at all the information through a new set of eyes. Initially Dublin Gulch was explored for other minerals, especially tungsten. After the Fort Knox deposit in Alaska was developed, everyone realized that the geological formation at Eagle was similar in nature. As Gray describes it, they looked at the ninety-five-million-year-old Cretaceous intrusions and said, "My God, it all runs 0.2 [grams per tonne]. If we mine enough of it, we can make a piss pot full of money."

According to Gray, the work done by Canada Tungsten when they took over the property was the best work that had ever been done at Dublin Gulch. The placer mining indicated that there was gold there. It had been mapped extensively, but many people walked over the granite without taking a second look. Gray's job in 2010 was to do all the legwork and make sure everything ran smoothly. But he took a non-conventional approach that John McConnell and the others came to appreciate, leading them to hire him in 2013.

When Paul Gray talks about the Eagle mine, his eyes brighten with a supercharged intensity. He knew the Eagle deposit was mineable from the New Millennium report in 1996, but the price of gold was too low at that time. The conventional business model was to define the ore deposit and sell the property or the company. But Victoria Gold decided instead to develop it from start to finish. Gray exclaims, "That's McConnell—always wanted to do it. A lot of juniors would have sold it or the assets, but McConnell always wanted to put it into production himself for the company, from the get-go.

"They have learned a lot from all the previous work. They could see mineralization at the contact boundaries, so they started looking for that zone. Geochemistry helped define that. They looked at all the old exploration—all of which falls along the same line. So they are now looking for intrusions along this line, mapping them and defining the margins, and that is the core of the exploration strategy today: to locate related zones of mineralization near the Eagle deposit, and flesh them out."

CHAPTER 5
THE EAGLE IS BORN

As spring turned to summer in 2017, the financing arrangement with BNP Paribas was finally consummated. On July 31, the company announced that two parties had executed a commitment letter for the debt-financing transaction previously outlined in January. BNP Paribas agreed, and obtained credit approval, to underwrite 100 percent of the planned US$220 million loan. McConnell noted that this deal would provide a significant chunk of the required project funding and left Victoria with the flexibility to source the remaining capital. The company was free to seek other ways to raise the balance of required project capital while minimizing dilution to existing shareholders.

In the margin:
In May of 2019, the haul roads were complete and the open pit was being readied for mining.

Buoyed with confidence that they could at last move forward with the construction, the company organized a formal shovel-turning ceremony at the Eagle Gold Project site on August 18, 2017. In front of a crowd of more than fifty invited guests, including politicians and members of the business community, John McConnell stated that since

Victoria Gold acquired the property in 2009, it had invested $130 million in studies, permitting and diamond drilling to get this far. Of that, $50 million went to Yukon companies. He also thanked Yukon premier Sandy Silver for lowering the taxes, making the Yukon a more favourable place to invest. McConnell noted that the Yukon had been good to him personally, and he had settled here with his wife and daughter.

He announced the next stage of development would be heap leach pad construction, and cut-and-fill of the crusher foundation. McConnell said that he hoped to have everybody back in two years' time to witness the first gold pour. First Nation of Na-Cho Nyäk Dun chief Simon Mervyn said he looked forward to working with all the contractors. "This is our land and we are happy to share it with you," he said. "We are not opposed to development, but we want to be part of the plan. And we will be, for years to come." Mervyn further characterized the relationship with Victoria Gold as one with a family. "You get along once in a while, we fight a lot, but usually we come to an agreement because we have respect for each other," he said.

Premier Sandy Silver said it was all about numbers, noting the significant capital investment that had gone into the project. "This is all good for the Yukon," he added, citing the partnership between Victoria Gold and the First Nation of Na-Cho Nyäk Dun as a model for First Nations in other jurisdictions. He thanked them for showcasing a partnership that would make the Yukon prosperous. Yukon member of parliament Larry Bagnell congratulated Victoria Gold for ten years of perseverance, noting that mining has been fundamental to the economy. He also acknowledged the leadership of what would become the largest corporate employer in the Yukon.

Special recognition for her contribution to the project was given to Sally Howson, who was retiring after many years of work. Howson had the longest-lasting connection with the project and the community, extending back twenty years. She was presented with handcrafted moccasins from Victoria Gold, and from the First Nation of Na-Cho Nyäk Dun, a lovely framed piece of local art crafted from fish scales and feathers. The ceremonial shovel-turning included McConnell, Chief Mervyn, Premier Silver and vice-presidents Mark Ayranto, Marty Rendall, Paul Gray and Tony George, who scooped up shovels of dirt from a pile in front of the assembled crowd while cameras clicked away. The construction of the Eagle Mine was about to begin.

Left: Sally Howson is recognized for her longstanding contribution to the project by John McConnell (left) and Chief Simon Mervyn.

Below: Shovel-turning ceremony at the Eagle Gold site on August 18, 2017. Left to right: Marty Rendall, Premier Sandy Silver, John McConnell, Chief Simon Mervyn and Mark Ayranto.

By November 2017, Victoria Gold could report on the completion of a wide range of preparatory activities—a program that was phrased to soothe prospective investors as "de-risking." The work included heap leach embankment preparation, control pond construction, upgrades to the access road and bridge, and road-building to the sites of the future crusher facilities and the gold recovery plant. Adding the camp units purchased the year before, Victoria upgraded the camp facility to a 250-bed capacity, expanded the size of the kitchen and dining area. Construction of the control pond would be completed by the end of the month. The securing of competitive prices on long-lead items began, and advanced detailed engineering was now 50 percent complete.[212]

The 250-person camp in April 2017.

There was more eventful news. On November 10, 2017, Victoria Gold and the Yukon Energy Corporation signed an agreement for the supply of electricity from the Mayo grid. The implementation of the agreement was dependent upon Victoria Gold raising the money through a final fundraising initiative. The new power line to the mine, which would be financed by Victoria Gold to the sum of eleven million dollars, would generate an additional ten million dollars a year over the ten-year life of the mine, and would benefit other Yukoners by a decrease in hydro rates of 1 to 2 percent. The improvements would include the construction of a new substation. Yukon Energy pointed out that the highest demand for electricity would be in the summer, when the demand from other customers would be lower.

The one troublesome issue that remained at this point was the financing. The BNP Paribas deal would only get Victoria halfway to securing the support to take the mine through to completion. According to McConnell, "Traditionally, a mine would be financed by debt and equity, i.e. you borrow money from the banks and you go out to capital markets and raise equity by selling Victoria stock. Normally it would be a sixty-forty split. We worked very closely with BNP Paribas and moved that forward to the point where they said, 'Here are the terms,' and we signed letters of engagement and that was done. But the capital markets are not kind right now so we couldn't raise the equity.

"The debt was there, but we had to find an alternative, so we approached a number of royalty companies… that buy a royalty so they get a percentage of every ounce you produce.

We approached them, and we eventually got it down to a couple of companies we were talking to and then the company we settled on was called Osisko Gold Royalties Ltd.

"So we had the royalty piece in place but one of the problems with selling a royalty is that it cannibalizes the debt. Because now the banks say you're getting a smaller share of income, so now we can only loan you *X*, which still left us short. We had talked to a firm called Orion Mine Finance. They are a very large hedge fund and they have a good relationship with Osisko, owning 19.5 percent of Osisko. We went back to them in late November of 2017, and we worked very closely with them for a month and closely with the Osisko guys. We are probably not on the BNP Christmas card list any more, but we came to what we thought was a far better financing package than the one we announced earlier in the year."

Critical to this phase was Marty Rendall, the chief financial officer for Victoria Gold. McConnell describes Rendall's role in the financing for the mine this way: "Marty and I have worked together for three different companies for almost thirty years. I'm a mining engineer, technical, but he has the financial expertise. So when we are doing project financing, he is very much front and centre. When we were meeting with investment banks, and there would be ten bankers in the room, I'd be there just to listen, watch body language, and Marty would take the lead in negotiations. I'd feel very comfortable at the end of the day because overall, he was generally the smartest guy in the room."

Marty Rendall, Victoria Gold's CFO.

Marty Rendall grew up in a small farming community in rural Ontario. After he graduated from Brock University with a degree in business administration, he took up the offer made by René Galipeau, a family friend, to come work for him in the mining sector. That was with Breakwater Resources, and he worked for the company for seven or eight years. One of their properties was the Nanisivik mine on Baffin Island in the high Arctic. That's where he first met McConnell, who was the mine manager at the time. Next he moved on to De Beers, which was opening diamond mines in the Northwest Territories.

Once again he found himself working with McConnell, who was in charge of developing the Snap Lake mine.

It was McConnell who approached Rendall about working for Victoria back in 2008. It was a big shift from working with a large company to a junior mining company with no track record. Rendall admits that it was both scary and exciting to sign on with a small outfit like Victoria, saying goodbye to the security and good pay at De Beers. But it appealed to his entrepreneurial spirit. For the first couple of months, he worked out of his basement, but eventually Victoria Gold got their own office, which they outfitted with used furniture.[213]

Rendall was looking for a change, and with the Nevada properties owned by Victoria Resources, he thought they would turn their attention to the south. But then Victoria acquired StrataGold and had to decide where to focus their attention. According to Rendall, it was the Eagle deposit at Dublin Gulch: "We really thought we could make a mine out of it," Rendall says. "It was back in 2009. It was early in the season, and it was looking good, and we believed in it. We believed it could be a mine. I think our assets in Nevada were at a little earlier stage. They were pretty sexy, but it was still questionable that you had a mine there. There was still a lot of work to do, a lot of money, as there was on Dublin Gulch. But I think when it came down to it, we said, 'Which one of these do we think we can turn into a real profitable mine?' And the answer was Eagle."

Investors were scared off by developing a mine in the north; they were apprehensive about the technicalities in working in a far northern environment. Could they do heap leach in the north? Many skeptics felt that it was not possible, but the Victoria team, with its collective northern experience, knew better. They opted for Dublin Gulch and sold their properties in Nevada. The Nevada properties were sold at a good time, because the market declined between 2013 and 2015. "We didn't have to go to the market to raise any money," recalls Rendall. "We had the money we got from the asset sale, which in retrospect, saved our bacon." Rendall stuck with Victoria Gold through some major ups and downs, but in the end he played a major role in the financing deal that ensured the construction of the mine.

But the press releases and congratulatory public announcements hid the tension and drama that was taking place behind the scenes. "The market really didn't understand the deal [struck with Osisko and Orion]," recalls Tara Christie. "The equity was well above market and we were trading at thirty-three cents when the equity was at fifty cents. It was the biggest deal the TSX [Venture Exchange] had had in—how long? Twenty-five years?[214]

"And yet people were negative about it, partly because they didn't understand it, and the markets were still really tough. And so it can't be good; they must've had to sell the farm. Well they didn't sell the farm. They actually got a great deal well above market. When you go through all the years of permitting and social licence from the First Nations and working with the community, these things actually stalemate if you don't actually get it financed and miss the cycle. It's usually ten or twenty years before somebody will pick it up or finance it again.

"As mining people, John and his team's reputations were on the line. Everybody had these expectations. And so that's an outside pressure—the money side and the financing are just one part of the story. The emotional roller-coaster of getting all those forces moving in the same direction is really an important part of the story. You need that social licence; if you hadn't financed it and had to wait another five years, permitting would have to be redone and the First Nation would have been unhappy. Back to the drawing board, and usually a new management team has to come in. If this team hadn't made it happen, then it would have been a long time and the Yukon wouldn't have the prosperity and low unemployment it has right now.

"We saw some pretty tense times," continues Christie. "We had to cancel the BNP debt before the other investors would say yes. Those little details were part of the drama in December. There were lots of days when things seemed pretty bleak, but John got up the next day to figure out how to get past, over or around it." Negotiating for half a billion dollars is not easy, and Marty Rendall was in the thick of things. The "shit to fun ratio" for a junior mining company was not very good, he admits; getting the next dollar was *always* a challenge, but raising half a billion was a Herculean task. Rendall has gone through fifteen financing arrangements during his time with Victoria Gold, but the most difficult one of all was the biggest one.

If Victoria Gold had continued to negotiate with BNP Paribas without any other options, BNP would have taken advantage of the situation, and Victoria would not have gotten a very good deal. BNP kept asking for a little more, and then a little more again. Eventually they asked for too much, and the deal soured. As BNP was trying to increase their take, Victoria opened the door to private equity funding. While private financing is faster and less onerous, it usually demands a higher percentage in return, but they were eager to do a deal, so they started edging their percentage down.

These negotiations are really a game of chicken, where both parties push their advantage and wait to see who would blink first. Victoria and Orion later played the same adversarial game of push and shove, but in the end, both parties saw the advantages of the deal. The dialogue changed to "Let's see what we need to do to get the deal done." They shook hands by the end of 2017, but that was not the end of the story.

By the end of 2017, Victoria Gold had just about spent all their money. They had to put money down on advance orders for machinery, and had hired on a crew. It looked like they would run out entirely by the end of February 2018. It was critical to close a deal by the beginning of March. "John [McConnell] is about the coolest cat I know," says Rendall. "There is a lot of pressure in his job, and risks. He always had two steady hands on the wheel. But that was the first time I saw him sweat—that financing. I was feeling a lot of stress and John was as well. We all felt that one."

Everybody put a lot of hard work bringing that deal to a conclusion. They didn't know if it would work for them (countless promising prospects never reach the mine stage). Meanwhile, they had one hundred people working in the Yukon. They were dead certain that they had a mine, but every day there was that nagging feeling that the funding deal wouldn't be consummated. According to Rendall, toward the end of February there were a couple of weeks where every two days there would be another stumbling block that threatened to kill the deal.

It was a real roller coaster ride. There would be the highs when they overcame an obstacle, and then there would be lows when more stumbling blocks appeared. Rendall recalls he didn't have trouble sleeping despite the building stress. He would work all night on those clauses and phrases, making phone calls and fixing problems, and fall asleep from exhaustion. Meanwhile they were dealing with multiple players at this point. Internally, there were major stockholders and members on the company board who didn't think that building a mine was the way to go, but McConnell's hand-picked team was convinced that there was a mine there, and they aimed straight and true for that target.

"There are lots of good projects that never get built," says Rendall. "This could have been one of them." If they hadn't struck a deal when they did, the game would have been over. The team would have been disbanded after ten years of hard work. At best, a new team would have been brought in; at worst, the project was finished. It was critical to close the deal. "It was do or die in my mind," says Rendall.

It finally came down to the wire. Rendall remembers a conference call at two o'clock in the morning in early March 2018. Their adversaries were tied into the conference call from New York and other places by telephone. The negotiations had been long and hard for both parties, and now the financiers were trying to renegotiate the terms again. In the end, Rendall explains, you have to be able to walk away if the terms aren't right. It was a decision by the chairman of the board, Sean Harvey, that finally forced the issue.

T. Sean Harvey had been fascinated by finance since childhood. He remembers one year in school when he had to make a choice between an automotive shop course and typing as an elective. He reasoned that he could hire somebody to fix his car, but he was going to need typing. The other bonus to this decision: he shared the class with thirty girls! The following year, he chose economics as his elective; he found the teacher of the class to be inspirational. He spent eleven years in university, during which time he acquired a BA and an MA in economics from Carleton University, a degree in law from the University of Western Ontario and an MBA from the University of Toronto.

Along the way, he acquired the nickname "Mr. Leisure," and his father wondered if he was ever going to get a job. "I do everything in moderation," he says, "except moderation. I take carpe diem to a new level. Hard core, full bore." Over the years he served as CEO for several different corporations, but then opted to do board work instead. He contributes his experience and knowledge to these companies, and personally invests his own money in their future. As chairman of the board for Victoria Gold, Harvey doesn't get involved in the nitty-gritty, but he plays an important role. John McConnell

Sean Harvey, chairman of Victoria Gold's board of directors (with Mark Ayranto in foreground).

explains it as being like the owner of a hockey team. Harvey doesn't get involved in the day-to-day details (these rest in the hands of the general manager and his staff), but in the end is brought in on the major decisions.

Harvey recalls that they were down to the wire on the big financing deal. It was beyond the eleventh hour; it was 11:59:59, and it was now or never. Mindful of the fact that McConnell and Rendall would have to work with their partners if the deal was to go ahead, Harvey assumed the role of the heavy in the deal-making. He dug his heels in and told the money people that their attempts to renegotiate the terms of the agreement had gone too far. Victoria Gold wasn't willing to agree to new terms at the last minute. The deal had been set with a handshake, he told them, and if they weren't going to honour that, he said in very blunt language, the deal was off.

There were four Victoria Gold representatives sitting around the telephone in Toronto, including Harvey, McConnell and Rendall, when he stated this ultimatum. There was a sharp intake of breath and dead silence. If this deal didn't happen, the prospects of building a mine looked grim. They had worked for ten years to reach this moment. The Victoria team was put on hold and the clock ticked away for four long minutes before the money people came back on the line. This was a big bet for them, and the rewards could be substantial, but there were concerns about the build and the potential for cost overruns.

Finally, they had their answer: the deal was a go. Victoria Gold was going to become a mine after all. The money started to flow into the Victoria account, starting with $150 million the day the deal was closed and signed in early March. "That was one of the best days of my professional life," says Rendall, "The deal was finally done and you get to call the guys up at the site and tell them, 'Don't worry, we've got the money!'" Harvey had been involved in mining ventures around the world, but says that there was something special about having a new mine start up in his "own backyard," here in Canada.

"Probably the most nervous I have ever been in my life was when we hadn't closed the financing, but had started the construction and were spending at the rate of probably a million dollars a day," recalls McConnell. "Like all good businessmen, Osisko and Orion at the last minute were trying to extract another pound of flesh from us, but we held firm and got the deal that we shook hands on before Christmas."

When Victoria Gold decided to go with Orion, Marty Rendall had to make the difficult phone call to BNP to inform them that they were out. He had been working with their people for more than a year and respected them. Rendall speaks with great respect about his financial adversaries: "They are all smart people doing the best deal they can for their stakeholders. They will try to get every penny they can out of their counterparts, while Victoria is trying to do the same thing. We try to get the best deal we can at the expense of our funding partners."

Together with an arrangement already made with Caterpillar Financial Services to purchase the mining fleet from Finning, the deal with Osisko and Orion amounted to more than five hundred million dollars to finance Victoria Gold through construction of the mine. Harvey is optimistic about the Eagle Mine venture. He feels this mine could last for fifty years. The eleven-year life of the mine was calculated with gold at a price of $1,250 an ounce. At $1,400 they can expand the pit and add another six or seven years. John McConnell is optimistic that the mine will last longer than currently projected, but remains more cautious. In his mind, it is good for at least twenty-five years. And that doesn't factor in the potential of the nearby deposits, like Shamrock and Nugget.

Chief Mervyn of the First Nation of Na-Cho Nyäk Dun was also pleased with the financing deal: "We have worked with Victoria for many years establishing a mutually beneficial relationship and they have exemplified the 'model' for companies wishing to work in our traditional territory. Several of our citizens are already employed early in the construction phase and many of our business partners are delivering value via construction contracts including earthworks, camp maintenance, catering, fuel supply, concrete and tire supply for the main mining equipment."[215]

Yukon premier Sandy Silver responded positively to the announcement of the deal: "We are very pleased to see the Eagle Gold Project take this important step forward. Today's announcement is great news for the territory and testament to the hard work of so many that have worked to make Yukon one of the world's most attractive places to invest. I would like to congratulate all parties including the First Nation of Na-Cho Nyäk Dun and Victoria Gold for getting the project to this stage in the process."[216]

As soon as the deal was announced, Victoria Gold moved quickly. Full construction commenced March 15, 2018, with completion and first gold pour scheduled for the second half of 2019. JDS Energy and Mining, which had the contract to manage the construction, got the green light on March 9, and by March 15 they were on-site and starting to open camp and begin work. Summit Camps, the camp catering contractor, worked together with the Na-Cho Nyäk Dun Development Corporation to do the camp expansion as a joint venture.[217] Victoria Gold wanted camp expansion from 250 beds to 450 completed by the end of April, so they bought a camp in Fort Nelson and had those beds on-site by April 1.[218] They weren't set up, but at least they didn't have trucks destroying the road, which was still not in the best condition, when the spring thaw began.

Pelly Construction had a major contract to undertake the earthworks associated with the mine development. At its peak, Pelly had more than 150 people on payroll, and as many as 75 being on-site at any time, when shift work was factored in. To keep things rolling, they used hot shifts, where the outgoing worker stepped down from the earth-moving equipment and the incoming shift climbed aboard without ever shutting down.

The most prominent of the earthworks was the construction of the embankment for the heap leach pad, commencing in the late summer of 2017. The work was planned all through the winter, and a lot of the preliminary work was done—some roads built, underground drainage and so forth—but then the financing was delayed. So they parked the equipment for the winter in anticipation of financing.

The embankment for the heap leach facility, May 2018.

Previous spread: The enlarged construction camp, May 2018.

When it came through, and Pelly showed up at site on March 15, they found that the two creeks they'd parked their equipment between had glaciered over, and the equipment was frozen in more than a metre of ice. They spent the first two weeks with jackhammers freeing the equipment before they could start work. But that wasn't the only problem that Pelly had to deal with. It was a very wet spring, so wet in fact that they couldn't get their servicing vehicles close enough to the equipment for maintenance work. As Keith Byram, president of Pelly Construction, recalls, "We were servicing our equipment from the back of a four-by-four... instead of having hoses that delivered the grease and oil to the equipment, we were carrying it in five-gallon pails. It was a nightmare. I don't think I've seen conditions so bad. And besides all this ice, it rained."

Eventually, the weather cleared up. Dry conditions in July, August and September allowed them to get back on schedule. But weather wasn't their only problem. Across the valley from the heap leach pad, they had to drill and blast a lot of rock. This work was also delayed through the winter because of financing. Finally, they started bringing the rock slope down where the crusher facilities were being installed, but they kept having slides, and had to go back to re-slope it several times. These problems, and others, created big headaches for Pelly. "I figure it was one of the worst summers we've ever had in our history, with all these problems," stated Byram later.

"There's always challenges," says Kevin Mather, president of JDS. "There are the known risks and the unknown risks, and you just have to find ways to push through it. That's the fun part of the work." At one time they had a geotechnical failure right above the two crushers. It started along a failure plane, so they flew a geotechnical expert from the United States up in a jet on a Friday night, and by Monday morning he had a plan and recommendations to move forward. The mining inspector approved it, and they got back to work. They had to redesign some walls, but all the buildings and infrastructure stayed where they were. "We might have lost a week on it," says Mather, "and if you don't move quick, you can easily lose a month."

Kevin Mather, president of JDS Energy and Mining.

By May 2018, engineering was 65 percent complete, with attention focused on Issued for Construction (IFC) drawings for concrete site works, power transmission and delivery of equipment. Fifty million dollars had already been spent, and another two hundred

Construction of the secondary/tertiary crushers and support walls.

million dollars committed. The decision was made to purchase all new major mobile equipment from Finning to improve operational efficiency and take advantage of attractive pricing. This decision and inflationary increases had raised the construction cost from $411 million to $442 million, but all major long-lead items had been procured, materially reducing the risk of schedule delays for construction.

Major earthworks were well under way on-site and were focused on the crushing facilities, gold recovery plant and heap leach facility. Concrete contracts had been awarded and casting of foundations for the crushing and gold recovery facilities was to start in July 2018. Camp expansion from 250 beds to full 450-bed capacity was to be completed by the end of May 2018. Total on-site construction work had already passed 185,000 hours without a lost time accident; emphasis was placed upon hiring from within the Yukon. Submission of the revised heap leach facility was deemed by the Yukon Water Board to have triggered an amendment to the water use licence. The company felt that all the required technical information had already been compiled, and was busy preparing an amendment application.[219]

Some of the huge Cat 785D 150-ton haul trucks purchased from Finning.

One of the two Cat 6040 shovels nicknamed "Beauty and the Beast."

Construction of the primary crusher in the background and the secondary/tertiary crushers in the foreground.

The base of the primary crusher.

The work proceeded quickly and smoothly. By December 2018, the construction was 60 percent complete. The earthworks were finished, and the concrete completion was at 90 percent. Assembly and installation of structural steel was one-third complete. A few weeks before, while the temperatures were still moderate for the time of season, the mine site was teeming with activity. Heavy earth-moving equipment working on the heap leach embankment resembled the activity on an anthill. Teams working for twenty-five different contractors were busy at the crushing and gold-processing facilities and finishing the heap leach pad. With 450 people working at the site, Dublin Gulch had become the sixth-largest community in the Yukon. In all, nearly a million hours of labour had been expended, with no lost time accident for six hundred thousand hours.

Summit Camps employees serve plenty of good food in camp.

The Eagle camp was running twenty-four hours a day during construction in December 2018.

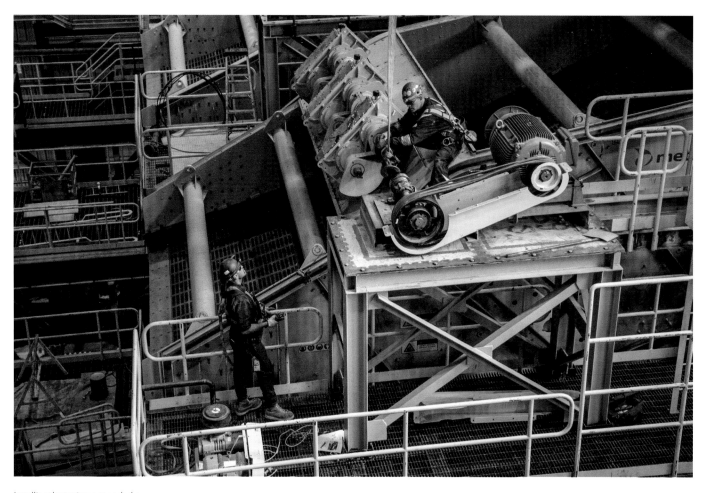

Installing the tertiary screen decks.

Interior of the secondary/tertiary crusher building.

Construction and assembly of the conveyor system.

Constructing the reclaim tunnel for the coarse ore conveyor to transport crushed ore from the primary crusher to the secondary/tertiary crushers.

The primary crusher and the secondary/tertiary crusher facilities, along with the fully constructed
coarse ore conveyor system connecting them.

Previous spread: View of the overland conveyor that carries crushed ore
from the secondary/tertiary crushers (the blue building in the distance)
to the heap leach facility. All crushed ore is moved by conveyor on-site.

Installing the geosynthetic clay liner at the heap leach facility, June 2019.

Installing the first of ten huge carbon columns in the gold recovery plant.

The camp, offices, workshops and events pond were all complete; installation of the electrical infrastructure, power lines and substation were progressing smoothly. One five-megawatt backup generator was already on-site and would be operational within three months, while others were being shipped to Dublin Gulch.[220] Staff recruitment was proceeding nicely, with the management team in place and the supervisors and key operators being recruited. Within months, more than a hundred employees would be on the Victoria Gold payroll. Many of them would be Yukoners.

More good news was announced. The most recent resource estimate resulted in an increase in the measured and indicated gold by 12.5 percent, while the grade of the reserve had increased by nearly 2.5 percent.[221] Construction progressed steadily as the new year of 2019 turned into spring. As work proceeded, preparations were made for commissioning of the twenty-two primary systems and seventy-nine secondary sub-systems necessary to operate the mine. Hiring continued, with 164 of the 260 mine staff hired by the beginning of May. By that time, the power line and distribution system on-site were ready for electrification.

The open pit was being readied for mining, and the haul roads were complete. The project had passed the million-hour mark in March 2019 without a lost time incident. The record remained proudly unbroken at 1.25 million hours. Optimism grew about the project potentially completing ahead of schedule.[222]

By June 2019, the construction of the mine was 95 percent complete. The primary crusher and conveyor systems were being commissioned and construction of the secondary/tertiary crushers was nearly finished. The commissioning of the project had advanced to 30 percent completion. The electrical supply from the Yukon Energy grid to the mine site was energized and a backup diesel electrical generation plant had been completed at a price of $10.5 million. Backup generators would come into use should there be a failure in the grid supply for any reason.

The original schedule to have the first ore delivered to the heap leach facility in August 2019 was confidently advanced to July, and a first gold pour was confirmed for September. By advancing the schedule, 2.5 million tonnes of ore was expected to be delivered to the heap leach facility by year end. All of this was accomplished with a proud safety record of 1.9 million person hours without a lost time incident. Staffing was progressing well and 50 percent of the two hundred staff already hired were recruited from within the Yukon.[223]

Previous spread: Welding the carbon columns in the gold recovery plant.

Construction of the gold recovery plant completed and proudly displaying the "Eagle Gold Mine" sign for the first time.

By the middle of July, many of the systems were commissioned and operating. High on the hillside overlooking the camp facility were the crushers on one side of Dublin Gulch and the gold recovery plant on the other. Preparations were being made to bring the pit into production. Blasting had commenced, and the fractured rock was being transported down to the primary crusher, where coarser material was being piled onto the heap leach facility. Half a million tonnes of ore had already been moved. Finer material was being stockpiled until pad preparations were complete and the first two lifts of coarse material had been laid down.

The Eagle substation and backup diesel generators at the Eagle Gold Mine site.

Previous spread: By May, the haul roads were complete
and the open pit was being readied for mining.

The McQuesten Substation near the Silver Trail, with the majestic Mount Haldane in the background.

Following spread: Aerial view of the Eagle Mine site. The primary and secondary/tertiary buildings to the right, the heap leach facility and gold recovery plant to the left.

Delivering explosives to the production holes in the open pit.

Right: Blasting is high-precision work. The production holes are filled with explosives, which are detonated according to a specific sequence.

Previous spread: The Caterpillar drill rig drilling production holes in the open pit.

Victoria Gold is proud that 25 percent of the Eagle Gold Mine workforce are women. Sonja Wilkinson is one of the many female haul-truck operators, delivering blasted ore to the primary crusher.

Above: A Cat 6040 shovel loading blasted ore into a Cat 785D haul truck.

Right: Delivering blasted ore from the open pit to the dump pocket of the primary crusher. View of the secondary/tertiary crushers in the right foreground. Crushed ore is placed on the lined heap leach facility, which is visible in the distance at the right-hand edge of the photo. The blue building immediately to the left of it is the gold recovery plant.

Above: Grasshopper conveyors delivering crushed ore onto the heap leach facility.

The crushers were busily grinding ore to the specified size; the noise was deafening, and ear protection was required. Crushing of the ore had begun in early July and material was being stockpiled. The conveyor system, which had the most complex controls of anything operating at the mine, was transporting the ore across the valley to the heap leach facility. The crushed ore in the first two lifts laid at the base of the heap leach facility was of a larger granular size to facilitate healthy percolation of the cyanide solution to the collection sump. The first layer of ore would be crushed to a size of thirty-eight millimetres, while the next layer above it was reduced to sixteen millimetres. The crushed ore in the third "lift" would be reduced to a standard 6.5 millimetres.

Testing the ore grade in the on-site assay lab.

The finishing details were being dealt with in the gold recovery plant in preparation for the first infusion of pregnant cyanide solution to begin working its way through the complex recovery process. The number of workers had declined significantly as the completion of construction approached. By mid-August, only fifteen members of the construction crew remained on-site, attending to the final details and aiding in the commissioning process as the mine moved into production.

In an interview in the *Whitehorse Star* on August 16, 2019, John McConnell spoke enthusiastically about the opening of the mine. He gave credit to the large workforce that had made the moment possible and spoke with pride about the number of Yukoners involved in the project. Half the workforce at Victoria Gold were territorial residents. A quarter were women, and 10 percent of the company payroll consisted of citizens of the First Nation of Na-Cho Nyäk Dun.

Following spread: A weak cyanide solution is applied to the crushed ore, which dissolves the gold from the ore. The gold bearing, or "pregnant," solution collects at the base of the heap leach facility. The pregnant solution is pumped via the large pipes from the base of the heap leach facility to the gold recovery plant.

There was great anticipation as the day of the first gold pour approached. The event was to be broadcast live at the Denver Gold Forum, an internationally recognized event taking place in Colorado in mid-September. McConnell was optimistic about the future of the mine. The price of gold had soared to fifteen hundred dollars per ounce, and with the exchange rate favouring the Canadian dollar, that figure was approaching two thousand dollars per ounce.

In a CBC interview with McConnell in early September, he had an opportunity to reflect on the long road to reach this point: "Financial markets have sucked over the years, making it difficult to raise the money to continue with the exploration and the development of the property, but we worked our way through that, and were patient, and we saw a financing window in 2018, and we took advantage of it."[224] When asked about the environmental impact of using a heap leach process that employs cyanide, he was clear in responding: the environment comes first. A detection system was installed to identify any leaks that might develop in the pad, which was constructed from four layers of plastic liner. "In my case," added McConnell, "I take a very personal interest in protecting the environment. I live here. This is my backyard too."

John McConnell, president and CEO, at the Denver Gold Forum to live webcast the first gold pour worldwide.

It had been raining earlier in the day on September 17, but the sun broke through later in the afternoon as more than a hundred people assembled in the Miner's Daughter, a lounge on Main Street in Whitehorse. The crowd was energized and the conversations noisy and upbeat. Present that afternoon were company staff, contractors and businesses that had been involved in the project. The media was there to record the event along with politicians including Whitehorse mayor Dan Curtis and several members of the territorial legislature, past and present, from both sides of the house. After all, there had been a change in government during the ten years the mine was under development.

Several hundred kilometres away, vice-president of operations and mine manager Dave Rouleau and Premier Sandy Silver waited expectantly to be brought online. In the darkened room at the Miner's Daughter, Mark Ayranto brought the conversation to a halt as the moment approached when all three sites—Denver, Whitehorse and the Eagle mine on Dublin Gulch—were connected by satellite. The screens at each end of the lounge flickered and came to life. From 3,100 kilometres away in Denver, John McConnell appeared on the screen to welcome everybody to the simultaneous broadcast of the event.

On September 17, 2019, by live webcast to audiences in Whitehorse, Denver and around the world,
Yukon premier Sandy Silver poured the first gold at the Eagle Gold Mine.

He made a few brief comments, and thanked key players including Hugh Agro, who believed in the project and encouraged John to check it out a decade before. A two-minute video recap of the mine construction over the past year was shown before the live broadcast shifted to the Eagle Mine. Dave Rouleau, standing in front of the state-of-the-art induction furnace, welcomed the faraway audiences and introduced Premier Silver, who was hardly recognizable, garbed in a reflective silver suit to protect him from the intense heat (gold is heated to 1,064 degrees Celsius). His otherworldly appearance elicited chuckles from the Whitehorse crowd.

It was Silver's task to initiate the pour of the first doré bar. Silver gave the thumbs up and the pour commenced. The induction furnace slowly tilted forward and brilliant molten gold slowly poured into the first mould and cascaded into a second and third mould below—it was over in a few seconds. The result: 1,001 troy ounces of gold. The Whitehorse party broke into loud and hearty cheers far from the mine as the pour was completed.

Back in Whitehorse, Mark Ayranto concluded by thanking all the many people who together made the Eagle Gold Project a success (there were too many to name them all). He extended a special acknowledgement to the First Nation of Na-cho Nyäk Dun and to Sally Howson, who started working on the Dublin Gulch project in the 1990s and retired while working with Victoria Gold in 2017. "Sally epitomized what the rest of the team tries to bring: you fall in love with the project, the community and the people, and Sally did that." He also thanked the members of the "ten-year-plus" club—John McConnell, Sean Harvey, Hugh Coyle and Steve Wilbur, as well as himself. He noted that he used to joke that the Eagle Gold Project was a ten-year overnight success. "But I can't say that anymore," he quipped, "because it's been more than ten years now!"

After praising the team that brought this project to conclusion, Mark Ayranto added that Victoria Gold, led by John McConnell, Sean Harvey and Marty Rendall, had brokered the largest financing on the venture exchange in twenty-five years. The Eagle Mine was a half-billion-dollar bet that was about to pay off. He added with a hint of pride that there were a lot of detractors who said that the project simply couldn't be done, that it wouldn't go. "We just put our heads down. We believed in the project and didn't listen to the naysayers and had the tenacity to get it done—and here we are!" He proposed a toast: everyone hoisted their champagne glasses and in a single voice, shouted, "Cheers!" And with that, the celebration began in earnest. The Eagle had been born.

Mark Ayranto in Whitehorse at the live webcast of the first gold pour.

Vice-President of Operations Dave Rouleau and Premier Sandy Silver holding the first gold poured at the Eagle Gold Mine on September 17, 2019 — a 1,001-ounce gold bar.

Sally Howson celebrating Eagle's first gold pour in Whitehorse, September 17, 2019.

CHAPTER 6

CONCLUSION

At its peak during construction, the Eagle Gold Mine employed around 800 people. As many as 450 men and women were working on-site at any time, excavating and moving earth, pouring concrete and assembling steel.[225] Machinery was put together, electronics were made functional and equipment was tested. During this phase, Dublin Gulch had a population that rivalled several Yukon communities: Carcross, Carmacks and most importantly, Mayo. When in operation, it would still have the tenth-largest population in the territory.

The first gold pour yielded 31.152 kilograms.

But while it rivalled them in size, it was not a community in the same sense that these other places were. Men and women were brought in for their skills and knowledge. People from other parts of the territory, from other provinces and in some cases from other nations came and went, contributing their expertise to bringing this complex organism to life.

As it moves into its operational mode, the population at the mine is becoming smaller and more organized into functions: mine operations and technical services, maintenance for fixed and mobile equipment, process operations, health and safety, environment, site services, supply chain and information technology. For the next decade, the mountain will be blasted and rock will be moved at the rate of 29,500 tons per day, crushed, and stacked where the gold will be extracted in a series of complex chemical processes. For all the millions of dollars invested, the thousands of tons of equipment imported and the millions of hours of labour expended, the final product will be reduced to small bars of gold, weighing thousands of ounces, to be shipped from the mine on a regular basis.

The impact of the Eagle Gold Mine cannot be overstated. Over the entire time that Victoria Gold has had an interest in the mine, tens of millions of dollars have been invested in exploration and development, and hundreds of millions of dollars have been spent to build the operation, including 150 million dollars that went to Yukon-based contractors. Over two million days of labour were employed to bring the mine to full operation, with almost half of all the work done with Yukon labour. The territory could proudly boast the lowest unemployment rate in the country, thanks in part to the Eagle Mine.

The mine will produce enough gold that it will pay for itself in two and a half years. Over the life of the mine, the territory will receive $110 million in royalties and another $240 million in income tax.[226] Nearly four hundred jobs will remain to operate the mine; half of these will be Yukon residents. With the additional job opportunities available because of the Eagle mine, every Na-Cho Nyäk Dun citizen can now find employment.[227] These numbers are significant to the Yukon territory, which is trying to reduce its dependence upon federal transfer payments.

Keith Byram, president of Pelly Construction, has seen this all before: "There's got to be a lot of permanent jobs. During the life of the mine, you have to haul a lot of stuff in. You've got chemicals, you've got fuel, the power generation. You know, everybody shudders at the word Faro these days, because of the big cleanup, but Faro totally changed this territory. It provided roads that would never have been built, it provided jobs for many years, high-paying jobs. You know, back in the days when Faro was really going, Whitehorse was quite an interesting town. Every night, seven nights a week, there were at least four places where you could dance to live music.

"There was a 'B' train of ore concentrates that came through Whitehorse every twenty minutes. The 'B' trains were those double trailers with the biggest legal load in Canada. The Yukon would not be anywhere near as developed if it hadn't been for the

Previous spread: Aerial view of the Eagle Mine site. Left to right: The camp, the gold recovery plant, the heap leach facility, the primary and secondary/tertiary buildings and the haul road to the open pit.

Faro mine, and the money poured out of that mine like crazy for years and years… the impact on the Yukon was very significant. The same thing with Victoria Gold, you know; it's going to be an engine of wealth creation for quite a while." The impact of the mine is spilling over into other sectors. Air North, which hasn't had regular air service to Mayo since the early 1980s, has been able to re-establish that service. "We now provide scheduled service five days a week with multiple flights on some days," says Joe Sparling, president of Air North.

With the demand for labour, there have been a number of Yukoners who have returned to the Yukon, and as they upgrade their skills over time, the percentage of local hires is likely to creep up over 50 percent: "Previously, a lot of people were leaving the Yukon to Alberta or British Columbia to find work in certain sectors," says Ranj Pillai, the territorial minister for energy, mines and resources. "Now they have the opportunity, especially with First Nations men and women, to come back. Not only can they be here, but they have the opportunity to sustain their culture, because for many people, their culture is around their communities and their traditions. I've had people talk to me about being able to come home so that their children can be around their grandparents. In turn, the grandparents are providing them with the opportunity to learn their language."

Pillai takes it further: "It's a plus for us because it is a company that has so many Yukoners in leadership roles. The organization has such an understanding of the Yukon— the community and the culture of the territory. There is an orientation that doesn't need to happen. And of course, Tara Christie has such a long history in Dawson with her family in the Yukon… And there are an amazing number of Indigenous people working on the project."

Michelle Dawson-Beattie is one of those who has been able to come home. Born and raised in the Yukon, she spent a lot of time at Champagne, where her family is from. She grew up on the land with her great-grandfather Alex Van Bibber and her "grandma," Sue Van Bibber. She spent a lot of time in the bush, trapping and hunting, when she was young. After a series of jobs working for First Nations, she found herself working as a liaison officer for a firm out of Edmonton. It was a good experience that gave her a whole new perspective on the Yukon, but despite coming home whenever she could to see her horses (she had a dozen), she missed her Yukon family. "You can always have money," she says, "but you can't have more memories."

Dawson-Beattie had talked to John McConnell about working for Victoria Gold, but she had almost given up hope that she would receive a job offer. Typical of the McConnell style, there was no formal interview, just a number of conversations. Finally, an offer letter came through. She received it on a Thursday, signed it on a Friday, put in her two weeks' notice on the Monday and was soon on her way home to her new job, her home, her family and her horses. Not long after arriving back in the Yukon, two mares gave birth. She named one of the foals Eagle, in tribute to getting the job at Victoria Gold. "I took a picture and sent it to John, and said, 'Hey, I want to introduce you to Eagle,'" she relates. "'If it hadn't been for you and the Victoria Gold mine happening, I wouldn't have been able to come home.'"

Nearly a year later, Dawson-Beattie was still enthusiastic about her work with Victoria Gold. She lamented that the first gold pour was going to occur in the middle of the hunting season, but she wouldn't miss the pour for anything. Mark Ayranto is big on hunting, she says. He told her, "You'll never have to worry about getting time off to go hunting." She has no problems standing behind what Victoria Gold is doing. Environmental impacts are highly regulated at the mine. Safety is the number one aim, and the Comprehensive Benefits Agreement sets the bar high for mining/First Nation relationships. Other mining companies are going to have to lift the bar even higher. "I want the First Nation to take hold of this great opportunity and use it to the best of their ability," she says. There are dozens more stories like hers.

NND Cobalt Mine Services is a family-owned operation (its predecessor, Golden Hill Ventures, operated in the Yukon for three decades). Now 51 percent owned by the Na-Cho Nyäk Dun Development Corporation, Cobalt undertook ancillary earthworks, construction of the mechanically stabilized earth (MSE) wall behind the primary crusher and installation of the McQuesten River Bridge. At its peak, it employed sixty, more than half of whom were First Nation of Na-Cho Nyäk Dun citizens. Cobalt has given back to the community, supporting Victoria Gold's Every Student, Every Day charitable initiative as well as local sports and cultural events.[228]

Rod Adams took a job operating a gravel crusher in the Yukon in 1984. Three years later, he bought the crusher and established Nuway Crushing. Starting with seven employees, Adams grew the company, which thirty years later crushes 90 percent of the gravel in the Yukon. Its operation at the Eagle Mine was entirely made up of Indigenous workers. Nuway has also given back to the community: over the past twenty years, the company has donated over $1.5 million to such charities as the Yukon Hospital Foundation or to sporting groups such as the Whitehorse Minor Hockey Association, the Klondike Road Relay and Special Olympics Yukon.[229]

Members of the Eagle Gold Mine Emergency Response Team (ERT).

The Na-Cho Nyäk Dun Development Corporation, which is headquartered in Whitehorse, was established in 1997 to create economic and employment opportunities for the local Indigenous people. The corporation pursues opportunities that benefit the First Nation and its citizens. The corporation, which is owned by the Na-Cho Nyäk Dun Business Trust, seeks opportunities to invest in or manage businesses active within their traditional territory. It forms partnerships, joint ventures and preferred contractor agreements to provide goods and services to customers like Victoria Gold.

Kluane Drilling has been responsible for most of the exploration drilling undertaken for Victoria Gold since 2009. Established over twenty-five years ago, the family-owned company specializes in diamond drilling with portable drills. The company has expanded until it now operates in countries around the world, but the headquarters remain in the Yukon. At the same time, it is proud of its involvement in many local charities including Every Student, Every Day. There are more stories like these.

There is another benefit to developing local businesses. As other big projects around the territory proceed to the construction phase, there are now local contractors with the skills and experience to make it happen. In fact, there are young people on student visas coming to study at Yukon College who are staying to work in the mining sector. The

Yukon has gone through boom-and-bust cycles before, but the construction of a new mine encourages optimism for the future of the territory.

Everybody on the management team expresses pride in the safety record of the company: a million and a half hours of labour without a lost time incident. Kevin Mather of JDS is quick to point out that the only incident to interrupt their accident-free record was one where the employee could have returned to work on light duty without a break in employment. The Yukon has had a reputation of being a hazardous place to work; there has always been a pervasive culture of cavalier disregard for safety in the Yukon workplace. Victoria Gold has set the bar high and has challenged the rest of the industry to emulate its success.

But what is the legacy that the Eagle Gold Mine and Victoria Gold will leave for the territory? It has already enhanced the economic well-being of the Yukon through contracting and employment, and will continue to do so during the next ten or more years of operation. Other sectors of the economy, such as construction, transportation and energy production, have already felt the positive economic benefits. The project has stimulated several new companies to develop expertise that they can carry forward to other projects in the future, and many Yukoners are improving their skill levels so they can move into more responsible and highly paid positions.

Ranj Pillai notes, "They have pulled together a number of Yukon companies that became the core of a mine build. We are now in a position where we have BMC and the Kudz Ze Kayah project, and Newmont Corporation with the Coffee project on the horizon. Many decades ago, the Yukon had a workforce and the technical capability to be able to come in and do these things. Now a group of companies have been brought together that are looking toward the next big project."

Whereas the previous mines in the region have led to improved transportation networks and infrastructure like a community hospital, the legacy of Eagle is likely to consist of more intangibles. One of the major beneficiaries of the project has been the First Nation. Not only does the First Nation of Na-Cho Nyäk Dun enjoy full employment and substantial contract work, but the relationship established between the company and the First Nation has also set an example for others to emulate. Things were different forty years ago; Indigenous people were still under the administration of the Department of Indian Affairs.

"It was tough," says Simon Mervyn. "Our people were living in shacks, and our elders were having a hard time because basically they had just stepped out of the tree line, if you know what I mean? They were living in tents in the wintertime, just barely living on scraps that the miners had left over. We had to beg with the Department of Indian Affairs for an old pickup and all that free stuff." The Indigenous people of Mayo were experiencing severe social disruption. Alcoholism was a major problem. Land claims changed all of that. Initiated in 1973 when the First Nations of the Yukon presented Prime Minister Pierre Trudeau with the document titled *Together Today for Our Children Tomorrow*, the negotiations to establish a land claim settlement continued for twenty years.

Both the umbrella final agreement with the Council for Yukon Indians (CYI) and the First Nation of Na-Cho Nyäk Dun final agreement were signed May 29, 1993. Among other things, the First Nation of Na-Cho Nyäk Dun received self-government, 4,739.68 square kilometres of settlement land and a cash settlement of more than fourteen million dollars, which was placed in a trust. For the first time since the European invasion during the gold rush, the First Nation was in a position to chart its own course. Now Indigenous people are in the game, not sitting on the sideline watching their natural resources being developed.

Much credit must be given to Victoria Gold and the leadership of John McConnell for establishing a positive relationship with the First Nation of Na-Cho Nyäk Dun. For its part, the First Nation took a progressive approach to establishing a relationship as well. It wanted more than jobs and training opportunities; it wanted in on the action. According to their legal counsel, Daryn Leas, the First Nation of Na-Cho Nyäk Dun came to the negotiating table wanting to share the profits of the venture. As Chief Simon Mervyn expressed during the 2017 shovel-turning ceremony, the Na-Cho Nyäk Dun were happy to share their traditional territory for development, provided adequate environmental safeguards were put in place, but he cast his glance toward the future, a future in which the First Nation played a part and shared in the benefits.[230]

This approach took Victoria Gold by surprise. There was no legal precedent or legal requirement for Victoria Gold to share the wealth, but to their credit, the company established a different way of looking at things. The desire to make the relationship work came from the top down. The process began more than twenty years ago when Sally Howson liaised with the First Nation, but it continued with Victoria Gold. It took two and a half years of negotiation to complete the Comprehensive Benefits Agreement. The First Nation of Na-Cho Nyäk Dun negotiated for jobs, training and contract opportunities, and pressed hard for the protection of the environment.

There was a full generation when First Nations of the Yukon were engaged in land claim negotiations. Things are different now; the First Nation of Na-Cho Nyäk Dun

Chief Simon Mervyn (left) and Mark Ayranto at the Comprehensive Co-operation
Benefit Agreement signing ceremony in Mayo on October 17, 2011.

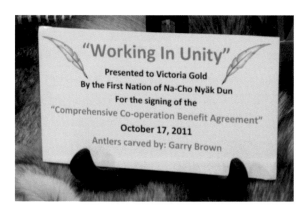

are actively managing their own affairs. They are not going to have some bank manage their money for them; they are doing it for themselves. They have developed a business acumen that enables them to engage in business conversations and negotiations in a way that was not possible fifty years ago. Now they share a common language with the business sector.

Another legacy of the Eagle Mine will be education. First there is the Victoria Gold Yukon Student Encouragement Society, Victoria's charitable initiative that works with communities to raise awareness and funds to support increased student attendance in schools throughout the Yukon. Established in the fall of 2012, the volunteer, not-for-profit Society has guided its Every Student, Every Day program in raising hundreds of thousands of dollars to support a variety of grassroots student attendance projects in the territory's schools and communities. Each day a student is absent from school diminishes students' chances to succeed later. With each missed day, they fall further and further behind.

The Society believes that making school attendance a priority helps Yukon students get better grades, develop healthy life habits, avoid dangerous behaviour and have an improved chance of graduating from high school. "By supporting increased student attendance through Every Student, Every Day," says John McConnell, "we are setting Yukon students up for a strong future. We are helping them feel more connected to their community, develop important social skills and friendships, and of course, be more likely to graduate from high school." Victoria Gold wants all Yukoners to be able to access employment opportunities at the new mine. For this to happen, however, children must attend school regularly and stay in class until they complete high school. "The Yukon needs every single soul now," says Debra Ryan, manager of strategic planning and alliances for Air North. "In today's workforce, whether it is the airline business or the mining sector, these young people have to have a certain amount of education."

Wendy Tayler, president and CEO of Alkan Air, sees that staying in school will naturally enable more Yukon students to pursue an advanced education: "Like John and Tara, I'm very passionate about education. They have Every Student, Every Day and I'm the chair of the Yukon Imagination Library, and also on the board of the college foundation. So, you bridge all of those together from birth and go all the way up through university, and it will naturally enable people to pursue advanced education. Victoria Gold will leave behind economic prosperity—for the community of Mayo, for the community of Dawson. For people who want to stay in the territory, they will be able to get those advanced jobs here at home. The fact that Yukon College is becoming a university means that they will be able to get their schooling at home."

Perhaps the real legacy is going to come five, six or ten years from now when some of the young people are graduating from school, and there are opportunities that they might not have had ten years ago. It is different now than it was a generation ago—now you can look to the future and see opportunities, not closed doors. Chief Mervyn also sees the opportunity in the future to incorporate traditional values into the school curriculum. So anything seems possible these days, perhaps even a school operated by the First Nation.

The development of the Eagle Mine is the result of several coincidental circumstances. In addition to having a mineable ore body, other factors have played into the equation for Victoria Gold. First, the extensive exploration work conducted by Canada Tungsten during the early 1980s attracted the interest of the mining fraternity. Second, StrataGold exercised the foresight to assemble a large block of claims under one ownership. The downturn in the economy in 2008 enabled Victoria Gold to acquire the Dublin Gulch property. An upturn in commodity prices a few years later enabled Victoria Gold to sell other holdings at advantageous prices, providing the much-needed capital to fund further exploration work in the vicinity of the Eagle Mine.

The success of the current venture can be attributed in part to the players who have been assembled to make the Eagle Gold Project reach completion. Many of them have been together, working toward completion, for more than ten years. The fact that they have stuck it out through good times and tight markets was critical to moving the project forward. Teamwork has been the key to completing such a complex and large-scale project, the largest and most expensive mine ever built in the territory.

But more than anything else, the success of the Eagle Mine can be credited to the determination and stick-to-it-iveness of one man: John McConnell. He's not your typical mining executive in a three-piece suit in a Bay Street office. He's more hands-on, unpretentious and approachable. "I didn't like working for a big company," he recalls. "De Beers was very frustrating, very much run out of South Africa, very bureaucratic. I would say the breaking point was one month when I travelled to Johannesburg from Yellowknife three times."[231]

He seems to be the polar opposite of a high-level mining executive, and though he has to travel around the globe, marketing his project to investors in all the major banking centres, he seems happier in a hard hat than in a suit, walking the talk rather than being settled in an office. "I like to be part of what's going on. Ultimately, I'm responsible for it, so I'd like to make sure it's done right."[232]

When his mine was in the planning stage, he moved to Whitehorse to be closer to the action. There, he could meet important decision-makers in the supermarket as well as the boardroom. That seemed to suit his personality. His meetings were as likely to take place over coffee or lunch as they were to take place in an office. This is borne out by the experiences of some of his closest associates at Victoria Gold, who seem to have been hired over coffee rather than through formal interviews. And he has stuck with his team, through thick and thin, from the earliest days of the company. There is an air of loyalty around the players working closest to him.

With his years of technical experience in the mining field, he was able to assess the viability of the Eagle deposit from a practical perspective. From the very beginning, he seemed to be confident of the potential and determined to make it happen. Victoria Gold had some great opportunities because of the timing of certain upturns and downturns in the market, but it wouldn't have happened if McConnell hadn't had perseverance and a true compass pointing to the mine.

McConnell has weathered challenging moments with an air of assured confidence. When speaking about the financial package he put together, and the massive expenditures he was committing to before the signatures of the financial deal were on paper, he makes it seem as if it were nothing at all. When he talks about stressful times when he was getting very little sleep, he dismisses it as par for the course. In fact, he directs the conversation with pride to issues like local employment, bringing Yukoners back to the Yukon, workplace safety and benefits to the territory. He is quick to praise the role that other members of his team have played in making the project succeed. Having an outgoing partner like Tara, a professional engineer born into the mining life, has made them the perfect power couple in Yukon mining.

Where does the mine go from here? It has a projected life of eleven years based on current proven reserves. Will that lifespan be extended like it was at the Fort Knox mine in Alaska? To answer that question, you must look up to the hills surrounding the Eagle Mine. One afternoon in July 2019, Paul Gray took me on a tour of the exploratory program Victoria Gold was conducting in the hills surrounding Dublin Gulch. The 4x4 truck climbed steadily up the steep, rocky trail into the hills above the Eagle Mine complex, toward the Nugget exploration camp several kilometres away.

It is a difficult road, more suitable for mountain goats than humans. I wouldn't want to attempt such an ascent in rainy weather, or when it was snowing. The sky was cloudless and sunny, and the air was hot. As we crawled along the sinuous track, Gray stopped at a high point in the Potato Hills. From there, we could look to the east and see the abandoned silver camp of Elsa in the distance. Shifting a little to the north, he pointed to McQuesten Lake, and to the mining village that was optimistically named Keno City a century ago. The distant hills receded into a smoky haze.

Gray pointed to a tiny white speck on a distant hill and explained that it was our destination, but it would be a roundabout journey to get there. The truck jostled and bounced, swerved and jolted along the primitive track cut into the hill—which could be a track to the future. We arrived at a long three-metre-deep trench newly cut into the hillside by a large Caterpillar excavator. Gray climbed down into the cut, stopping here and there to examine the rocks, occasionally cracking one open with his rock hammer. He was looking for a seam of green rock indicating a mineral contact zone with ancient sediments that they are trying to trace across the hilly terrain. Farther on, he pointed out a trench from which grab samples of soil had been carefully collected and bagged, in preparation for shipping to an analytical lab. He stopped at a small spring to check the water level, because it is critical to their continuing work and camp operation.

On a rocky outcrop of a nearby peak, we stopped again to visit a diamond drill operated by Kluane Drilling that was boring a hole hundreds of metres into the hillside to extract a long cylindrical column, or core, of rock. The cores are boxed and carefully logged and sent off for analysis. The work is hot and sweaty, noisy (ear protection is required) and repetitive as the cores are brought to the surface and boxed. We returned to Nugget camp, a cluster of neatly arranged white canvas tents, and after a drink of iced tea, we hiked up to a small shed where the samples are carefully logged and prepared. It's all in the details, Gray explained as he showed me where data is entered in a computer. Meticulous records are kept of every sample—precise location, depth and so on—so that there is no questioning the reliability of laboratory results.

Every drill core is split in half using a diamond saw, so that one half of each sample can be retained for future reference. When the results come back, they are entered into a modelling program from which a three-dimensional image can be built up of a potential ore body. If it is large enough, it can be mined. I asked the big question: "Is there more mineable rock out here? Is there more of a future for Victoria Gold's Eagle Mine?" Gray responded enthusiastically: "There's too much smoke; this is a mature, emerging area. It's here somewhere!" That is the confidence that makes mines.

Looking back more than a hundred years, numerous brave and optimistic men scoured these hills looking for an exposed vein that could be developed into a mine.

Living in isolated, challenging conditions, they blasted and hacked their way into the hillsides for hundreds of metres, following their dreams. Little did they know that this rock would someday develop into the largest gold mine the Yukon has ever had. Hidden within the rocky matrix of these hills is microscopic gold in minute quantities that the older technologies were incapable of making economical.

Paul Gray inspecting core samples at Nugget camp.

But with today's technologies, millions of ounces of gold trapped within this bedrock can be released. If only people like Bobbie Fisher, Stewart and Catto, Carscallen and Sprague had lived long enough to see their dreams fulfilled. I wonder what they would have thought if we could transport them from the past to witness the scope of the current mine. Would they have been shocked, amazed or satisfied that their optimism has been realized? Could they possibly have comprehended the advances in technology that allow for aviation, satellite communication and digital technology?

When Letha MacLachlan visited the prospective mine several years ago, she only learned later that her ancestor had worked the same area with horse and pack saddle a century before. And what a difference a hundred years had made. The early prospectors worked under more challenging conditions. There was no internet or satellite communication. There were no helicopters or roads capable of carrying large trucks. Back then it was all done with backpack or pack saddle. The men searching for gold might be out of touch for weeks at a time without contacting another soul. There were campfires rather than camp kitchens; sleeping bags rather than bunkhouses. No DEET to keep away the clouds of mosquitoes that were known to drive humans mad.

But then and now, they all shared the same dream: that hidden somewhere amid the craggy outcrops on remote hillsides, they would find the motherlode that would make them wealthy. During the intervening century, men and women continued to prospect, probe and test, gather samples and seek investors to back their efforts, hopeful of making that next big discovery. Families lived here and grew up on the creeks. Transportation and communications improved, analytical methods advanced and new, more efficient recovery systems evolved that turned a dream into reality—that, and the determination of all those who sought the pot at the end of the rainbow, the ones who dared to dream and think big and turn those dreams into reality.

That dream would become the Eagle Mine.

Endnotes

1 Roots in Bleiler et al. 2006: 10.

2 First Nation of Na-Cho Nyäk Dun 2002: iv.

3 Greer 1995: 7.

4 For more detail on early mining in the Yukon valley, see Gates 1994.

5 McKim 1920: 28.

6 Mayo Historical Society 1990: 36.

7 Mayo Historical Society 1990: 458–59; "The Dublin Hydraulics, Ltd." *Dawson Daily News,* December 22, 1910: 17; Aho 2006: 50 states he arrived in 1898, which would have been impossible if he had been involved in the Spanish-American War.

8 "Suttles Brings in Good Mining News," *Dawson Daily News,* November 16, 1910: 4; see also Aho 2006: 51.

9 "Mining Gossip" KMYA September 17, 1898: 2.

10 Aho 2006: 51.

11 McKim 1920: 28.

12 McKim 1920: 28.

13 Mayo Historical Society 1990: 42–43.

14 Mayo Historical Society 1990: 56–57.

15 "Haggart Creek Is Examined," *Yukon Morning World,* June 10, 1908: 4.

16 Bostock 1990: 159–61.

17 "The Dublin Hydraulics, Ltd," *Dawson Daily News,* December 22, 1910: 17.

18 Keele 1904 from Bostock 1957: 137.

19 Frank Taylor personal communication, May 12, 2019. The left limit is the left side of the valley when looking downstream.

20 "Report on the Outlying Placers," *Dawson Daily News*, April 5, 1909: 3.

21 Keele 1904 from Bostock 1957: 137.

22 "Dublin Gulch for Hydraulics," *Yukon World,* January 10, 1905: 1.

23 "Dublin is Serene," *Dawson Daily News,* August 5, 1908: 4.

24 "Public Notice," *Dawson Daily News,* April 3, 1909: 3; "Organize Company," *Dawson Daily News,* April 24, 1909: 3.

25 "The Dublin Hydraulics, Ltd," *Dawson Daily News,* December 22, 1910: 17.

26 "Placer Mining Concession to Be Thrown Open," *Whitehorse Star,* February 16, 1917: 1; "Interesting Happenings in the Northern End of Territory," *Whitehorse Star,* March 2, 1917:4; *The Victoria Daily Times,* November 18, 1909: 5 states that the hydraulic company was actively working the concession in 1909.

27 Mayo Historical Society 1990: 458.

28 "Dublin Gulch Case to Be Through Today," *Dawson Daily News,* October 20, 1914: 4; "Dublin Gulch Case Has Been Settled," *Dawson Daily News,* October 23, 1914: 4.

29 Mayo Historical Society 1990: 41.

30 Mayo Historical Society 1990: 235, 458.

31 Gates 2017: 48–61.

32 "Would Develop a New Industry in the Yukon," *The Ottawa Journal*, July 30, 1917: 12.

33 Aho 2006: 51–52.

34 "The Towns Where Silver Was King," *Yukon News*, September 2, 2016, http://yukon-news.com/letters-opinions/the-towns-where-silver-was-king/

35 Mayo Historical Society 1990: 235–37.

36 Frank Taylor personal communication, May 12, 2019.

37 Ibid.

38 Fred Taylor interview by Tara Christie, March 15, 22 & 24, 2015.

39 Mayo Historical Society 1990: 236.

40 Bostock 1939: 8; Bostock 1990: 159–61.

41 Fred Taylor personal communication, May 12, 2019.

42 Bostock 1939: 8.

43 Fred Taylor interview by Tara Christie, March 15, 22 & 24, 2015.

44 Bostock 1941: 15.

45 Bostock 1940: 6–7.

46 Fred Taylor interview by Tara Christie, March 15, 22 & 24, 2015.

47 Mayo Historical Society 1990: 372.

48 "Govt. Will Prospect for Scheelite in YT," *Nanaimo Daily News*, July 8, 1942: 6.

49 "Man Seriously Injured," *Dawson Daily News*, September 17, 1940: 2; "Young Miner on Dublin Gulch Is Seriously Injured," *Whitehorse Star*, October 4, 1940: 1.

50 Mayo Historical Society 1990: 41.

51 Bostock 1990: 191–92, 194.

52 "Will Explore Area in Mayo District," *Dawson Daily News*, June 17, 1943: 1.

53 "Mayo Camp Now Shipping Scheelite," *Dawson Daily News*, September 17, 1942: 1.

54 Bostock 1990: 196–97.

55 "Mayo Camp Now Shipping Scheelite," *Dawson Daily News*, September 17, 1942: 1.

56 In the latter part of the nineteenth century and the first quarter of the twentieth century, the Yukon was an isolated location and difficult to reach, especially in the winter when the major tributaries were frozen. People adopted the term "outside" to refer to any place beyond the Yukon— for example, Vancouver or Seattle. The term is still in popular use today.

57 Frank Taylor interview by Tara Christie, March 15, 22 & 24, 2015.

58 Frank Taylor interview by Gates, April 4, 2019

59 Debicki 1983: 44.

60 Frank Taylor interview by Tara Christie, March 15, 22 & 24, 2015.

61 Debicki 1983: 82.

62 Debicki 1983: 104; Aho 2006: 61.

63 Mayo Historical Society 1990: 41.

64 Aho 2006: 56.

65 Debicki 1983: 111; Aho 2006: 56.

66 Aho 2006: 56.

67 Mayo Historical Society 1990: 338–39; Aho 2006: 57.

68 Howard White correspondence to Gates, July 23, 2019.

69 Aho 2006: 57.

70 Frank Taylor interview, April 4, 2019.

71 Holway interview, March 1, 2018.

72 Gutrath, n.d.: 8.

73 Lennan personal communication, November 23, 2019.

74 Department of Indian Affairs and Northern Development 1990: 34.

75 Department of Indian Affairs and Northern Development 1993: 35–36; 1996: 102–4.

76 Department of Indian Affairs and Northern Development 1998: 142–44; 2003: 167.

77 For example: Department of Indian Affairs and Northern Development 1992: 35–36; 1993: 35–37; 1996:1 02–4; 1998: 142–44; 2003: 167.

78 McKim 1920: 28.

79 Yukon Archives GOV 0278; See also McKim 1920.

80 Cathro 2006: 113.

81 Yukon Archives GOV 0278; See also McKim 1920.

82 Gibbons 1911.

83 Yukon Archives GOV 0278.

84 Yukon Archives GOV 0278.

85 Mayo Historical Society 1990: 372.

86 "Quartz Groups of Robt. Fisher in Mayo Camp," *Dawson Daily News*, November 29, 1920: 11.

87 For his involvement in the Lone Star mine, Dr. Catto was inducted into the Yukon Prospectors' Hall of Fame. See https://www.yukonprospectors.ca/ypa_site_003.htm#halloffame

88 Yukon Archives GOV 0278; McKim 1920: 28; "Robert Fisher, One of Extensive Owners in Mayo Silver Area," *Dawson Daily News*, December 12, 1921: 35.

89 Mayo Historical Society 1990: 354; Smyth 1991: 12; "Frank Carscallen of Yukon Dead," *The Vancouver Province*, April 3, 1936: 13.

90 "Yukon Putting Money in Quartz," *Dawson Daily News*, April 21, 1910: 4.

91 "Stewart on the Outlook," *Dawson Daily News*, February 13, 1911: 4.

92 "Rich Quartz Now on Dublin Gulch," *Dawson Daily News*, August 29, 1911: 4.

93 "Mills of the North," *Dawson Daily News*, September 4, 1911: 2.

94 "Good Reports from Dublin Country," *Dawson Daily News*, September 4, 1911: 4.

95 "Quartz Man is in the City from Dublin." *Dawson Daily News*, December 14, 1911: 4.

96 Mayo Historical Society 1990: 454; "Mrs. Annie Hall and B.C. Sprague Cupid's Victims," *Dawson Daily News*, December 27, 1911: 4.

97 Letha MacLachlan email communication to Mike Burke, July 11, 2017.

98 MacLean 1914: 128.

99 MacLean 1914: 143–44.

100 MacLean 1914: 150.

101 MacLean 1914: 158.

102 "Yukon Quartz Claim is Sold for Large Sum," *Dawson Daily News*, February 28, 1913: 4.

103 Ibid; Mayo Historical Society, 1990: 60–62, 425.

104 "To Work on Quartz on Dublin Gulch," *Dawson Daily News*, November 7, 1914: 4.

105 "Government Assayer Reports on His Work for a Year," *Dawson Daily News*, March 16, 1914: 4.

106 "Use Diamond Drill," *Dawson Daily News*, February 26, 1914: 2.

107 Mayo Historical Society 1990: 57.

108 "Work is Under Way in Mayo Camp," *Dawson Daily News*, November 4, 1915: 4.

109 *Polk's Yukon Gazeteer and Directory* (1915–16): 665–66.

110 "Tungsten is Being Staked in the Yukon," *Dawson Daily News*, January 8, 1916: 4.

111 Lennan 1983: 245.

112 "Would Develop a New Industry in the Yukon," *The Ottawa Journal*, July 30, 1917: 12.

113 Lennan 1979: 6; Bostock, H.S. "Special War Minerals in the Yukon," *Dawson Weekly News*, April 17, 1942: 3.

114 "Expert Needed," *Dawson Daily News*, October 16, 1917: 2.

115 "Robert Fisher, One of Extensive Owners in Mayo Silver Area," *Dawson Daily News*, December 12, 1921: 35.

116 "Cattle for Mayo," *Whitehorse Star*, August 26, 1921: 1; "First Cattle Arrive in Mayo," *Whitehorse Star*, September 23, 1921: 1; "Cattle Killed on Top of Keno Hill," *Dawson Daily News*, September 24, 1921: 4.

117 "Meetings in Mayo Camp," *Dawson Daily News*, November 21, 1921: 4.

118 Smyth 1991: 15; "Entire Vote in from Mayo District," *Dawson Daily News*, December 13, 1921: 4.

119 "Notice of Grant of a Poll," *Dawson Daily News*, November 28, 1921: 3.

120 "Byng Had a Wonderful Trip Over Keno Hill," *Dawson Daily News*, August 11, 1922: 4.

121 "Extensive Silver and Gold Mining in the Klondike District," *Dawson Daily News*, November 29, 1920: 16.

122 McKim 1920: 28.

123 "Robert Fisher, One of Extensive Owners in Mayo Silver Area." *Dawson Daily News*. December 12, 1921: 35.

124 "Robert Fisher, One of Extensive Owners in Mayo Silver Area," *Dawson Daily News*, December 12, 1921: 35.

125 "Fisher Properties in the Keno Hill Area Have Fine Showing," *Dawson Daily News*, November 11, 1922: 34.

126 "Bob Henderson Returns from Prospecting," *Dawson Daily News*, September 9, 1922: 4.

127 "Reports Big Tungsten Find on Dublin Gulch," *Dawson Daily News*, August 13, 1919: 1.

128 "Former Dawson Man Dies in the State of Virginia," *Dawson Daily News*, November 3, 1925: 4.

129 SRK Consulting 2008: 4.2.

130 Bostock 1940: 8.

131 Yukon Geological Survey's Integrated Data System, http://data.geology.gov.yk.ca/Occurrence/13910; Debicki 1983: 16–17.

132 Anonymous, n.d., http://data.geology.gov.yk.ca/Occurrence/1391

133 SRK Consulting 2008: 4.2.

134 Debicki 1983: 108, 114, 120.

135 Lennan 1984: 245.

136 "Govt. Will Prospect for Scheelite in YT," *Nanaimo Daily News*, July 8, 1942: 6; "Survey Parties Find Deposits of War Metals," *Winnipeg Tribune*, December 30, 1942: 12.

137 "New Mineral Discoveries Stimulate Yukon Interest," *The Edmonton Journal*, June 5, 1943: 17.

138 Bostock 1990: 191–94; Mayo Historical Society 1990: 41.

139 "New Mineral Discoveries Stimulate Yukon Interest," *The Edmonton Journal*, June 5, 1943: 17.

140 Aho 2006: 57.

141 SRK Consulting 2008: 4.2.

142 Lennan 1984: 245, http://data.geology. gov.yk.ca/occurrence/15059

143 SRK Consulting 2008: 4.2.

144 Lennan personal communication, November 23, 2019.

145 SRK Consulting 2008: 4.2; Nordin 1981.

146 SRK Consulting 2008: 4.2.

147 Yukon Geological Survey's Integrated Data System, http://data.geology.gov. yk.ca/Occurrence/13910

148 Ibid; In "Friedland Play Combines Gold, Oil" (*National Post BC Report*, May 13, 1995), the Dublin Gulch property is described as a "Fort Knox look-alike."

149 "Project Description for the Fort Knox Mine," report prepared by Fairbanks Gold Mining, Inc., Fairbanks, AK, 1992.

150 "Amax Gold Closes Fort Knox Deal," *Canada News-Wire*, January 7, 1992; "Amax Inc. Unit Swaps Stock for Two Firms, Alaska Gold Reserve," *The Wall Street Journal*, American edition, January 8, 1992.

151 "Amax Gold Discussing Buying Fairbanks Gold," *Wall Street Journal Europe*, August 28, 1991; "Amax Gold to Acquire Fairbanks Gold." *Canada News-Wire*, September 12, 1991.

152 "Friedland Play Combines Gold, Oil," *National Post BC Report*, May 13, 1995.

153 Quandt et al. 2008: 22–23.

154 Personal communication between Bob Williamson and Tara Christie, August 23, 2019.

155 "Yukon Exploration Activity Picking Up," *The Northern Miner*, May 10, 1993.

156 Yukon Geological Survey's Integrated Data System, http://data.geology.gov. yk.ca/Occurrence/13910

157 In "Friedland Play Combines Gold, Oil" (*National Post BC Report*, May 13, 1995), the transaction is described as a reverse take-over of Starmin Mining, a Toronto-listed shell company.

158 "Growth Forecast for First Dynasty," *The Northern Miner*, May 15, 1995.

159 "First Dynasty Mines Completes Financing," *The Northern Miner*, June 12, 1995.

160 JDS Energy and Mining 2016: 1–6.

161 "Growth Forecast for First Dynasty," *The Northern Miner*, May 15, 1995.

162 Precise numbers vary from one report to another, but the magnitudes of the numbers are similar. JDS Energy and Mining, 2016: 1–6; "First Dynasty Tallies Reserves for Dublin Gulch," *The Northern Miner*, March 25, 1996.

163 Yukon Geological Survey's Integrated Data System. http://data.geology.gov. yk.ca/Occurrence/13910

164 "First Dynasty Ups Dublin Gulch Reserves," *The Northern Miner*, February 10, 1997.

165 *Canadian American Mines Handbook* 2004–5: 397.

166 "Cornucopia Eyes Dublin Gulch," *The Northern Miner*, September 1, 1997; "Letter to the Editor—Activity at Dublin Gulch in Yukon Understated," *The Northern Miner*, September 22, 1997.

167 "First Dynasty to Retain New Millennium," *The Northern Miner*, November 10, 1997.

168 *Canadian Mines Handbook* 2003–4: 190; With the change of name, control of the company went to Twin Star, a private

company controlled by Indian industrialist Anil Agarwal. Robert Friedland's share in the company was reduced from 24 percent to 13.5 percent; "First Dynasty to Retain New Millennium," *The Northern Miner*, November 10, 1997.

169 Yukon Geological Survey's Integrated Data System, http://data.geology.gov.yk.ca/Occurrence/13910

170 "Kinross Completes Acquisition of Bema Gold Corporation," press release issued February 27, 2007, https://www.kinross.com/news-and-investors/news-releases/press-release-details/2007/Kinross-Completes-Acquisition-Of-Bema-Gold-Corporation/default.aspx

171 Langford and Russell 2018: 53.

172 "Eagle Gold Project Ready to Soar in Yukon," *Canadian Business Journal*, www.cbj.ca/victoria-gold/; McConnell interview, July 8, 2018; Chad Williams, the first CEO of Victoria after the Kinross acquisition, also takes credit for inviting John McConnell onto the board of directors, in Langford and Russell 2018: 53.

173 Christie and McConnell interview, July 8, 2018; "Eagle Gold Project Ready to Soar in Yukon," *Canadian Business Journal*, www.cbj.ca/victoria-gold/

174 Langford and Russell 2018: 53.

175 Q4 2009 Victoria Gold Financial Statement: 5, 11.

176 National Instrument 43–101 (the "NI 43-101" or the "NI") is a national instrument for the Standards of Disclosure for Mineral Projects within Canada. The instrument is a codified set of rules and guidelines for reporting and displaying information related to mineral properties owned by, or explored by, companies which report these results on stock exchanges within Canada. This includes foreign-owned mining entities who trade on stock exchanges overseen by the Canadian Securities Administrators, even if they only trade on over-the-counter (OTC) derivatives or other instrumented securities.

177 "Mining Explorers 2009: Victoria Gold Corporation." *North of 60 Mining News,* https://www.miningnewsnorth.com/story/2009/11/01/news/mining-explorers-2009-victoria-gold-corporation/2104.html

178 Starting out as a small trailer rental company in Calgary in 1947, ATCO specialized in providing portable industrial modular housing and innovative modular facilities. It has grown and diversified into other sectors including energy, transportation and real estate. With nearly six thousand employees worldwide, it is still recognized as a leader in portable modular housing.

179 Ayranto interview, September 13, 2018.

180 Victoria Gold Corp., "Victoria Gold Corp. Announces Positive Pre-feasibility Results for the Eagle Gold Project," press release issued March 9, 2010, https://www.vitgoldcorp.com/news/victoria-gold-corp-announces-positive-pre-feasibility-results-for-the-eagle-gold-project/

181 MacLean 1914, Chapter III.

182 The actual inscription reads *Olive Mineral Claim Dublin Gulch Y.T.*, Kinsey and Kinsey [photographers] 1912.

183 Victoria Gold Corp., "Victoria Gold Corp. Achieves Another Important Milestone at Eagle Gold Project, Yukon," press release issued December 20, 2010, https://www.vitgoldcorp.com/news/victoria-gold-corp-achieves-another-important-milestone-at-eagle-gold-project-yukon/

184 Victoria Gold Corp., "Victoria Gold Corp. and Na-cho Nyäk Dun First Nation Sign MOU on the Eagle Gold Project," press release issued May 20, 2010, https://www.vitgoldcorp.com/news/victoria-gold-corp-and-na-cho-nyak-dun-first-nation-sign-mou-on-the-eagle-gold-project/.

185 Langford and Russell 2018: 56.

186 Victoria Gold Corp., "Victoria Gold Corp. Announces Further Progress on the Path to Development of the Eagle Gold Mine," press release issued February 7, 2011, https://www.vitgoldcorp.com/news/victoria-gold-corp-announces-further-progress-on-the-path-to-development-of-the-eagle-gold-mine/

187 Victoria Gold Corp., "Victoria Gold Announces 2011 Dublin Gulch Exploration Program," Press release issued February 28, 2011, https://www.vitgoldcorp.com/news/victoria-gold-announces-2011-dublin-gulch-exploration-program/

188 Victoria Gold Corp., "The First Nation of Na-Cho Nyäk Dun and Victoria Extend Exploration Agreement at Dublin Gulch, Eagle Gold Deposit, Yukon," press release issued June 7, 2011, https://www.vitgoldcorp.com/news/the-first-nation-of-nacho-nyak-dun-and-victoria-extend-exploration-agreement-at-dublin-gulch-eagle-gold-deposit-yukon/

189 Victoria Gold Corp., "Victoria and First Nation of Na-Cho Nyäk Dun Sign Comprehensive Cooperation Benefits Agreement, Eagle Gold Deposit, Yukon," press release issued October 14, 2011, https://www.vitgoldcorp.com/news/victoria-first-nation-of-na-cho-nyak-dun-sign-comprehensive-cooperation-benefits-agreement-eagle-gold-deposit-yukon/; see also CBA Committee Reports for 2017 and 2018, for example.

190 Victoria Gold Corp., "Positive Feasibility Study Outlines Annual Gold Production of 200,000 Ounces at Victoria's Eagle Gold Project, Yukon," press release issued February 22, 2012, https://www.vitgoldcorp.com/news/positive-feasibility-study-outlines-annual-gold-production-of-200-000-ounces-at-victorias-eagle-gold-project-yukon/

191 Victoria Gold Corp., "Victoria Gold Sells Relief Canyon Property in Nevada," Press release issued March 27, 2012, https://www.vitgoldcorp.com/news/victoria-gold-sells-relief-canyon-property-in-nevada/

192 Victoria Gold Corp., "Victoria Gold Enters into Agreement to Sell the Cove Property in Nevada," press release issued April 10, 2012, https://www.vitgoldcorp.com/news/victoria-gold-enters-into-agreement-to-sell-the-cove-property-in-nevada/

193 Victoria Gold Corp., "Victoria Gold Enters into Agreement to Sell its Mill Canyon Property in Nevada," press release issued May 25, 2012, https://www.

vitgoldcorp.com/news/victoria-gold-enters-into-agreement-to-sell-its-mill-canyon-property-in-nevada/; "Victoria Provides Update on Nevada Asset Sales." Press release issued June 6, 2012. https://www.vitgoldcorp.com/news/victoria-gold-enters-into-agreement-to-sell-its-mill-canyon-property-in-nevada/.

194 Victoria Gold Corp., "Victoria Achieves Major Permitting Milestone for its Eagle Gold Project, Yukon," press release issued September 4, 2012, https://www.vitgoldcorp.com/news/victoria-achieves-major-permitting-milestone-for-its-eagle-gold-project-yukon/.

195 Victoria Gold Corp., "Victoria Gold and the First Nation of Na-Cho Nyäk Dun Sign Historic Access and Exploration Agreement on 290 km² of Virgin Land along Strike to Dublin Gulch, Yukon," press release issued September 13, 2012, https://www.vitgoldcorp.com/news/victoria-gold-the-first-nation-of-na-cho-nyak-dun-sign-historic-access-and-exploration-agreement-on-290-km2-of-virgin-land-alo/.

196 Victoria Gold Corp., "Victoria Identifies New Gold Targets on the Dublin Gulch Property, Yukon," press release issued January 27, 2013, https://www.vitgoldcorp.com/news/victoria-identifies-new-gold-targets-on-the-dublin-gulch-property-yukon/.

197 Victoria Gold Corp., "Victoria Extends Gold Mineralization at Olive, Adjacent to Victoria's Eagle Gold Deposit, Yukon," press release issued January 28, 2013, https://www.vitgoldcorp.com/news/victoria-extends-gold-mineralization-at-olive-adjacent-to-victorias-eagle-gold-deposit-yukon/.

198 Victoria Gold Corp., "Yukon Environmental Assessment Board Approves Victoria's Eagle Gold Project," press release issued February 19, 2013, https://www.vitgoldcorp.com/news/yukon-environmental-assessment-board-approves-victorias-eagle-gold-project/.

199 Victoria Gold Corp., "Victoria Achieves Major Milestone with Receipt of the Quartz Mining License for the Eagle Gold Project, Yukon," press release issued September 22, 2013, https://www.vitgoldcorp.com/news/victoria-achieves-major-milestone-with-receipt-of-the-quartz-mining-license-for-the-eagle-gold-project-yukon/.

200 Victoria Gold Corp., "Victoria Provides Status Report on Eagle Gold Project," press release issued May 6, 2013, https://www.vitgoldcorp.com/news/victoria-provides-status-report-on-eagle-gold-project/

201 Victoria Gold Corp., "Victoria Initiates Drilling to Test High Grade Target Adjacent to Eagle Project," press release issued May 20, 2014, https://www.vitgoldcorp.com/news/victoria-initiates-drilling-to-test-high-grade-target-adjacent-to-eagle-project/; "Victoria Completes Successful Exploration Program at High-Grade Olive Zone," press release issued November 4, 2014, https://www.vitgoldcorp.com/news/victoria-completes-successful-exploration-program-at-high-grade-olive-zone/;

"Victoria's High Grade Olive Zone is Heap Leachable," press release issued August 7, 2015, https://www.vitgoldcorp.com/news/victorias-high-grade-olive-zone-is-heap-leachable/.

202 Victoria Gold Corp., "Victoria Gold Receives Final Major Permit for Eagle Gold Project, Yukon," press release issued December 7, 2015, https://www.vitgoldcorp.com/news/victoria-gold-receives-final-major-permit-for-eagle-gold-project-yukon/.

203 Various press releases from 2016 record the results of the Victoria Gold drilling program, including February 23; March 31; May 5, 13 & 24; June 15 & 28; July 6 & 20; November 22. https://www.vitgoldcorp.com/news/#2016

204 The strip ratio refers to the ratio of the volume of or waste material handled in order to extract some tonnage of ore. The lower this number, the better.

205 JDS Energy and Mining 2016: 1.9, 21.1.

206 Victoria Gold Corp., "Victoria Gold Completes $28,778,750 Bought Deal Financing," press release issued August 31, 2016, https://www.vitgoldcorp.com/news/victoria-gold-completes-28-778-750-bought-deal-financing/.

207 Victoria Gold Corp., "Victoria Gold Engages Financial Advisor and Closes $4.7 Million Flow-Through Financing," press release issued November 17, 2016. https://www.vitgoldcorp.com/news/victoria-gold-engages-financial-advisor-and-closes-4.7-million-flow-through-financing/; Victoria Gold Corp., "Victoria Gold Reports on Olive-Shamrock 2016 Phase 2 Drill Program," press release issued November 22, 2016. https://www.vitgoldcorp.com/news/victoria-gold-reports-on-olive-shamrock-2016-phase-2-drill-program/; Victoria Gold Corp., "Victoria Gold Announces $6.2M 2017 Program to Unlock the Exploration Potential of the Dublin Gulch Property," press release issued January 19, 2017, https://www.vitgoldcorp.com/news/victoria-gold-announces-6.2m-2017-program-to-unlock-the-exploration-potential-of-the-dublin-gulch-property/.

208 Victoria Gold Corp., "Victoria Gold: US$220M Project Debt Facility for the Eagle Gold Mine," press release issued January 24, 2017, https://www.vitgoldcorp.com/news/victoria-gold-us-220m-project-debt-facility-for-the-eagle-gold-mine/.

209 Victoria Gold Corp., "Victoria Gold Awards Engineering for the Eagle Project to JDS/Hatch Team," press release issued March 27, 2017, https://www.vitgoldcorp.com/news/victoria-gold-awards-engineering-for-the-eagle-project-to-jds-hatch-team/.

210 Victoria Gold Corp. "Victoria Gold: Acquires Cat Mining Fleet for the Eagle Gold Project," press release issued March 28, 2017, https://www.vitgoldcorp.com/news/victoria-gold-acquires-cat-mining-fleet-for-the-eagle-gold-project/.

211 Gray interview, November 16, 2018.

212 Yukon Energy Corp., "Yukon Energy and Victoria Gold Sign Power Purchase Agreement," press release issued November 14, 2017, https://yukonenergy.

ca/about-us/news-events/yukon-energy-and-victoria-gold-sign-power-purchase-agreement.

213 Rendall interview, October 2, 2019.

214 The TSX Venture Exchange was established in 1999.

215 Victoria Gold Corp., "Victoria Gold Announces Execution of Definitive Documentation for the Eagle Financing Package and Closing of Equity and Royalty (First Tranche)," press release issued April 16, 2018, https://www.vitgoldcorp.com/news/victoria-gold-announces-execution-of-definitive-documentation-for-the-eagle-financing-package-and-closing-of-equity-and-royalty/

216 Ibid.

217 Mather interview, May 14, 2019.

218 Victoria Gold Corp., "Victoria Gold Provide, Eagle Development Update, Yukon Territory," press release issued May 24, 2018, https://www.vitgoldcorp.com/news/victoria-gold-provides-eagle-development-update-yukon-territory/

219 Victoria Gold Corp., "Victoria Gold Provides Eagle Development Update, Yukon Territory," press release issued, May 24, 2019. https://www.vitgoldcorp.com/news/victoria-gold-provides-eagle-development-update-yukon-territory/.

220 Victoria Gold Corp., "Victoria Gold: Eagle Mine Construction is 60% Complete." Press release issued December 4, 2018. https://www.vitgoldcorp.com/news/victoria-gold-eagle-mine-construction-is-60-complete/.

221 Victoria Gold Corp., "Victoria Gold: Eagle Measured & Indicated Resource Increases by 450,000 oz Au." Press release issued December 5, 2018. https://www.vitgoldcorp.com/news/victoria-gold-eagle-measured-indicated-resource-increases-by-450-000-oz-au/.

222 Victoria Gold Corp., "Victoria Gold: Eagle Mine Construction 90% Complete." Press release issued May 7, 2019. https://www.vitgoldcorp.com/news/victoria-gold-eagle-mine-construction-90-complete/.

223 Victoria Gold Corp., "Victoria Gold's Eagle Mine Nearing Operation," press release issued June 4, 2019, https://www.vitgoldcorp.com/news/victoria-golds-eagle-mine-nearing-operations/

224 CBC Yukon, "Victoria Gold's Eagle Gold Project... Yukon's largest gold mine," video: https://www.facebook.com/cbcyukon/videos/victoria-golds-eagle-gold-projectyukons-largest-gold-mine/382528425756104/

225 "Mine's First Gold Bar Pour Is Imminent," *Whitehorse Star,* August 16, 2019: 9.

226 Pillai interview, July 11, 2019.

227 Mervyn interview, June 17, 2019; Leas interview, May 15, 2019.

228 "Partners: NND Cobalt Mine Services," *Yukon News*, July 26, 2019: 10.

229 Hansard, Yukon Legislative Assembly, 34th legislature, 2nd session, No. 37; October 16, 2017: 1114; "Partners: Nuway Crushing," *Whitehorse Star*, May 24, 2019: 37.

230 Simon Mervyn at shovel-turning ceremony at Dublin Gulch, August 18, 2017.

231 Langford and Russell 2018: 51.

232 Langford and Russell 2018: 57.

Glossary

adit: An entrance to an underground mine that is horizontal or nearly horizontal, by which the mine can be entered, drained of water and ventilated, and by which minerals are extracted at the lowest convenient level. Adits are also used to explore for mineral veins.

adsorption/desorption gold recovery (ADR) plant: A mining facility where gold dissolved in a cyanide solution is recovered by adsorption using activated charcoal. The final loaded carbon is then removed and washed before undergoing desorption of gold cyanide at high temperature and pH. The elute solution, normally consisting of caustic soda (the electrolyte), cyanide and water, circulates through the loaded carbon, extracting gold and other metals. The gold is then removed from the solution by electrolysis. The solution is recycled through the loaded carbon, extracting more gold and other metals until all the gold has been removed.

assay: A chemical test performed on a sample of ores or minerals to determine the amount of valuable metals contained.

block caving: An inexpensive method of mining in which large blocks of ore are undercut, causing the ore to break or cave under its own weight.

cam drive: A cam is a rotating or sliding piece in a mechanical linkage used especially in transforming rotary motion into linear motion. In a specific mining application, this would impart a shaking movement to a **sluice** bed.

catskinner: A person who operates a Caterpillar tractor.

crusher: A plant that reduces the size of ore particles by mechanical means. At Dublin Gulch there are three crushers—a primary, a secondary and a tertiary—each of which progressively reduces the particulate size of the ore to 6.5 millimetres before the ore is deposited on the **heap leach pad**.

deadmen: Objects buried in or secured to the ground for the purpose of providing anchorage or leverage.

decommissioning a placer mine: According to Schedule 1 of the Yukon's Placer Mining Act, measures must be taken to return the site to a safe and stable condition when a placer mine ceases operation. The vegetative mat must be re-established, all equipment and metals must be removed, camps hauled away, slopes graded to a safe incline to minimize erosion and trenches backfilled.

diamond drill: A rotary type of rock drill that cuts a core of rock that is recovered in long cylindrical sections, two centimetres or more in diameter.

Discovery: When placer gold is found in a new stream not previously prospected, the first claim staked on that creek becomes known as the Discovery claim. All other claims staked on that creek are measured either upstream (above Discovery, or A/D) or downstream (below Discovery, or B/D) from the first claim. Typically, they are described thus, for example: No. 10 A/D or No. 27 B/D.

doré bar: A semi-pure gold brick, smelted at the site of a mine and then transported to a refinery for further purification. Gold doré bars have other elements, often silver, left in them.

dragline: An excavating machine in which the bucket is attached by cables to a long boom and operates by dragging the bucket toward the machine.

dredge: A floating mechanical gold excavator. Often several storeys high, the dredges referred to in this book consist of a continuous bucket line that excavates gold-bearing gravel deposits and dumps the gravels into a large cylindrical rotating screen, or trommel, inside the structure of the dredge. The gravel is drenched with water as this sloping drum rotates, and the finer particulates containing the gold fall through the perforations in the screen onto a series of **sluice boxes**, where the gold is captured. The remaining material, which has been washed clean, is carried along a boom or stacker that extends from the stern of the dredge, and deposits the gravel, now washed clean of gold, into the pond behind. As the dredge excavates the gravels at the front and deposits the tailings at the back, it slowly moves forward in its own pond.

dredgemaster: The foreman or lead hand controlling the operation of a dredge. Other members of a dredge crew include the bow decker, the stern decker, the oiler and the winchman. In the Yukon, dredges operated around the clock during the short summer season, run by a crew of three or four men per shift.

failure plane: The interface between two stratified layers of sedimentary or meta-sedimentary rock formations, where slippage or shearing can occur.

fine gold: There are two applications of this term in **placer mining**. In one, fineness is the proportion of pure gold or silver in jewellery or bullion expressed in parts per thousand. Thus, 925 fine gold indicates 925 parts out of 1,000, or 92.5 percent pure gold. In the other, the term refers to the size of the gold particles. The small particulate size of fine gold makes it harder to capture in a **sluice box**.

float: Pieces of rock that have been broken off and moved from their original location by natural forces such as frost, gravity or glacial action.

flow-through shares: In Canada, junior resource corporations often have difficulty raising capital to finance their exploration and development activities. Moreover, many are in a non-taxable position and do not need to deduct their resource expenses. The flow-through shares allow the issuing corporation to transfer the resource expenses (deductible for tax purposes) to the investor.

flume: An artificial channel, generally constructed of wood, conveying water.

galena: Lead sulphide, the most common ore mineral of lead. In the Elsa/Keno Hill area, the galena had a high silver content.

gold recovery plant: See **adsorption/desorption gold recovery plant**.

granite pluton: In geology, a pluton is a body of intrusive igneous rock that is crystallized from magma slowly cooling below the surface of the earth. Granite and granodiorite are among the most common rock types in a pluton.

grubstake: An early form of credit. In pioneer times before the discovery of the Klondike, traders along the Yukon valley would provide prospectors with a grubstake, essentially an advance of food and supplies against repayment from the future recovery of gold.

hardrock mining: A method of extracting valuable minerals from solid rock, either by tunnelling or excavating an open pit, as opposed to **placer mining**, which is the extracting of valuable minerals from weathered surface deposits of loose, unconsolidated gravels.

heap leach pad: An enclosure into which ore is placed for the purpose of dissolving gold from ore using a cyanide solution. The ore is crushed to the optimal granular size for maximum extraction of gold or other valuable minerals. The enclosure, or heap leach pad, is lined with several layers of impermeable membrane to prevent the loss of the gold-rich cyanide solution.

heap leach process: In low-grade gold bearing ore, the most efficient way to recover the gold is through the heap leach process. The ore is crushed and reduced to a small size (in the case of the Eagle deposit, the particle size is 6.5 millimetres) and placed in a specially prepared containment, or **heap leach pad**, which is lined with a membrane or membranes to prevent the loss of liquid. A cyanide solution is then percolated through the crushed ore, and the gold is dissolved and the pregnant solution (cyanide solution bearing dissolved gold) is captured and piped to the **adsorption/desorption gold recovery (ADR)** plant, where the gold is extracted from the solution.

hydraulic mining: In **placer mining**, gold is removed from unconsolidated gravels by washing the gravel with high volumes of water under pressure. As the lighter gravels are washed away, the denser gold is left behind and is then recovered by washing it through a **sluice box**.

lay: In **placer mining**, an owner of a claim might have a person or persons take a lay on the claim, allowing them to work the claim for the owner in exchange for a share of the gold recovered.

NI 43-101 resource estimate: A national instrument for the Standards of Disclosure for Mineral Projects within Canada. The instrument is a codified set of rules and guidelines for reporting and displaying information related to mineral properties owned by, or explored by, companies which report these results on stock exchanges within Canada.

outside: In the latter part of the nineteenth century and the first quarter of the twentieth century, the Yukon was an isolated location and difficult to reach, especially in the winter, when the river transportation was not possible. People adopted the term "outside" to refer to any place beyond the Yukon, for example Vancouver or Seattle. The term is still in popular use today.

patented claim: Mineral rights are usually obtained by staking a claim; however, minerals may also be held by private entities and originate from either Crown grants or patents or freehold tenures that were issued as part and parcel of another type of grant. The owner of such privately held minerals is entitled to conduct reconnaissance and exploration activities and develop those minerals, provided that he or she obtains the necessary surface access.

pay dirt: Loose, unconsolidated material containing enough gold to make mining profitable. Similar to a pay streak, which is a zone of placer deposits containing enough pay dirt to make mining it profitable.

placer mining, placer prospecting: The extraction of gold particles from loose, unconsolidated gravels in which the gold has been weathered out of the bedrock. The gold is separated from the gravel in which it is found using large volumes of water which are run through a sluice box, into which the gold-bearing gravel has been placed. The primary tools for placer prospecting are a shovel and a gold pan.

pregnant cyanide solution: See **cyanide heap leach process**.

prospecting lease: A licence granted by the government to an individual or individuals for the purpose of prospecting upon that land. The lease is granted for a year but may be extended for two additional years if at the end of each year the leaseholder is able to demonstrate that an expenditure has been made on the property as described in the lease. The lease may not be longer than five miles along any creek or stream, and is one thousand feet wide.

quartz claim/quartz prospecting: A quartz claim is a parcel of land located or granted for hard rock mining. A claim also includes any ditches or water rights used for mining the claim, and all other things belonging to or used in the working of the claim for mining purposes. Mineral tenure is granted under the free entry system in the Yukon. This system gives individuals the exclusive right to publicly owned mineral substances from the surface of their claim to an unlimited extension downward vertically from the boundary of the claim or lease. All **Commissioner's lands** are open for staking and mineral exploration unless they are expressly excluded or withdrawn by order-in-council.

quartz mining licence: A major **hardrock mining** project in the Yukon requires a detailed environmental and socio-economic assessment and various regulatory approvals. These approvals include but are not limited to a **water use licence** and a quartz mining licence. A project must go through two distinct stages before mining activity can commence. First, an assessment determines whether significant adverse environmental or socio-economic effects are likely to occur. Second, a regulatory licensing approval process needs to take place.

reverse-circulation drilling/drill holes: A method of drilling that uses dual wall drill rods that consist of an outer drill rod with an inner tube. These hollow inner tubes allow the drill cuttings to be transported back to the surface in a continuous, steady flow.

scheelite: A tungsten mineral with the chemical formula $CaWO_4$. In Dublin Gulch, it has been found in unconsolidated gravels and as hardrock deposits.

selling a royalty: When major financing for a gold mine is undertaken, some investors will provide such funds in exchange for a specified percentage of all the gold recovered.

shaft: A vertical or inclined excavation in rock for the purpose of providing access to an ore body. Usually equipped with a hoist at the top, which lowers and raises a conveyance for handling workers and materials.

sluice box: A long, narrow, open-ended box inclined slightly so that water will flow through it. Placer gravel is excavated and dumped into the upper end of the sluice box, where it is mixed with generous amounts of water. As this mixture flows down the incline, the gold falls to the bottom of the box and is trapped between rows of ribs, or riffles, that are typically placed in the bottom of the sluice box at right angles to the flow of water.

staking a claim: When a promising geological deposit is discovered by an individual, it is staked or marked out in a prescribed manner. This action is then filed with the mining recorder. Once the claim is granted, the claim holder is granted certain rights regarding the use of the property for mining purposes. Different conditions apply to **placer** and to **hardrock** claims. The size of the claim is prescribed by the government mining regulations and is subject to occasional change.

strip ratio: The ratio that compares the amount of waste rock that must be removed from an open-pit mine per tonne of ore that is removed. The lower the ratio, the more efficient the mining.

sulphide content: Sulphide minerals are the major source of world supplies of a wide range of metals and are the most important group of ore minerals. They are also potential sources of pollution. In particular, the release of sulphur through the weathering of sulphides in natural rocks or in mine wastes generates sulphuric acid, resulting in acid rock drainage.

tungsten: One of the hardest materials in the mineral world, with the chemical symbol W. It is very dense (three times heavier than iron) and has the highest melting point of any element in the periodic table. In Dublin Gulch it is found in hardrock as well as in placer deposits. Because its density matches that of gold, it is captured in **sluice boxes** along with gold when **placer mining**. Because of its density, high melting point and hardness, it can be used in bullets and other military applications. It is particularly useful in armour-piercing shells and in protective armour.

water control pond: The control pond contains all "contact water," i.e. all overflow water or water from mine facilities that is directed to this pond (by ditches and/or pipes). That water is either reused in the process (as make-up water for the **heap leach** facility) or discharged to the environment. Almost all the collected water is needed for the for the **heap leach** process, but before any water is discharged into the environment it is tested to ensure it meets water quality standards as defined in the **water use licence**.

water use licence: Under the Yukon Waters Act, the Yukon Water Board issues water licences for various activities for the use of water and/or the deposit of waste to water. The process is intended to promote the balance of conservation, development and utilization of Yukon water for all Yukoners and Canadians.

Selected Sources

Interviews by the Author

Mark Ayranto, September 13, 2018.

Keith Byram, June 30, 2019.

Tara Christie and John McConnell, July 8, 2018.

Janet Dickson, October 23, 2018.

Tony George, November 19, 2018.

Paul Gray, November 16, 2018.

Sean Harvey, October 3, 2019.

Ron Holway, March 1, 2018.

Sally Howson, February 26, 2019.

Daryn Leas, May 15, 2019.

Letha MacLachlan, QC, September 12, 2018.

Kevin Mather, May 14, 2019.

John McConnell, October 2, 2019.

Chief Simon Mervyn, June 17, 2019.

Hon. Ranj Pillai, July 11, 2019.

Marty Rendall, October 2, 2019.

Joe Sparling and Deb Ryan, May 22, 2019.

Wendy Tayler, May 22, 2019.

Frank Taylor, April 4, 2019.

The author attended and witnessed the August 18, 2017, shovel-turning ceremony at Dublin Gulch, including quotations from Simon Mervyn.

Mark Ayranto is also quoted from author's attendance at "first pour" celebration in Whitehorse, September 17, 2019.

Newspapers and Periodicals

Dawson Daily News

Dawson Weekly News

Edmonton Journal

Nanaimo Daily News

Northern Miner

Ottawa Daily Journal

Whitehorse Star

Published Sources

Aho, Dr. Aaro E. *Hills of Silver: The Yukon's Mighty Keno Hill Mine*. Madeira Park, BC: Harbour Publishing, 2006.

Bleiler, Lynette, Christopher Burn and Mark O'Donoghue. *Heart of the Yukon: A Natural and Cultural History of the Mayo Area*. Mayo, YT: Village of Mayo, 2006.

Bostock, Hugh S. "Mining History of the Yukon" 1935–1940. Canada, Department of Mines, Memoirs 193, 209, 218, 220, 234. Ottawa: King's Printer, 1936–41.

———. *Pack Horse Tracks*. Whitehorse: Yukon Geoscience Forum, 1990.

———. "Potential Mineral Resources of Yukon Territory" (Rev). Ottawa: Geological Survey of Canada, 1954.

———. "Results of Investigation of the Tungsten Deposits in Yukon." Unpublished report. Whitehorse: EMR Library, call number TN271.T8 B67, 1940.

———. "Yukon Territory; Selected Field Reports of the Geological Survey of Canada, 1898 to 1933." Geological Survey of Canada Memoir 284. Ottawa: Queen's Printer, 1957.

Cathro, Robert J. "Great Mining Camps of Canada 1: The History and Geology of the Keno Hill Silver Camp, Yukon Territory." *Geoscience Canada* vol. 33 no. 3 (September 2006): 103–34.

———. "Potato Hills Tungsten Mayo Area, Yukon Territory." Report prepared for Connaught Mines Ltd. Copy on file with Yukon Geological Survey, 1970. http://yma.gov.yk.ca/060841.pdf

CBA Committee. "CBA Committee Annual Implementation Report." 2017 & 2018 booklets, produced by Victoria Gold Corp.

Christie, Tara. "Telling Our Stories: Placer Miners 2015/2016, Interview and Lifestyle Research Summary." Report prepared for Tr'ondëk-Klondike World Heritage Committee, 2016.

Debicki, Ruth. *Yukon Mineral Industry 1941–1959*. Whitehorse: Department of Indian Affairs and Northern Development, 1983.

———. *Yukon Placer Mining Industry 1978–1982*. Whitehorse: Exploration and Geological Services, Northern Affairs Program, 1983.

Department of Indian Affairs and Northern Development. "Yukon Placer Mining Industry" 1985–88; 1991–92; 1993–94; 1995–97. Whitehorse: Placer Mining Section, Mineral Resources Directorate, 1990, 1992, 1993, 1996, 1998.

———. *Kluane First Nation Final Agreement*. Ottawa: Public Works and Government Services, 2003.

Doherty, J. Allan, and J.A. VanRanden. "1994 Assessment Report on the Len Property, Mayo Mining District, Yukon, August 23–24, 1994." Whitehorse: Report number 093253, EMR Library, 1994. http://yma.gov.yk.ca/093253.pdf

Gates, Michael. *From the Klondike to Berlin: The Yukon in World War I*. Madeira Park, BC: Harbour Publishing, 2017.

———. *Gold at Fortymile Creek: Early Days in the Yukon*. Vancouver: UBC Press, 1994.

George, Arnold. "Report on the Outlying Placers." *Dawson Daily News*, April 5, 1909: 3.

Germaine, Peter, Denise, and Cathy, *Stories of the Old Ways for the Future Generations*. Mayo, YT: First Nation of Na-Cho Nyäk Dun, 2001.

Gibbons, C.H. "Christie and the Bear." *Wide World Magazine* vol 26, no. 126 (April 1911): 388–94.

Greer, Sheila. "Archaeological and Historic Sites Impact Assessment, Dublin Gulch Mining Property Final Report." Edmonton: Report prepared for Hallam Knight Piésold Ltd. on behalf of First Dynasty Mines, 1995.

Gutrath, Gordon. "Gordon Gutrath—Yukon Experience." Unpublished manuscript.

———. "Preliminary Geological Report on the Mar Claim Group, Mayo Mining District, Yukon Territory." Report prepared on behalf of Queenstake Resources. Copy on file with Yukon Geological Survey, Whitehorse, 1978. http://yma.gov.yk.ca/090364.pdf

JDS Energy and Mining, Inc. NI 43-101 Feasibility Study Technical Report for the Eagle Gold Project, Yukon Territory, Canada. Report prepared for Victoria Gold Corp., Vancouver, 2016.

Keele, Joseph. "The Duncan Creek Mining District (Stewart River, Yukon Territory)," 1904 in Bostock, *Yukon Territory, Selected Field Reports of the Geological Survey of Canada 1898 to 1923*, GSC Memoir 284, 1957: 127–43.

Langford, Cooper, and Rhiannon Russell. "Ten Years. Two Mines." *Up Here Business* (Fall 2018): 49–57.

Lennan, W.B. "1978 Project Report on the Dublin Gulch Property Yukon Territory Mayo Mining District." Report 090471, filed with the Yukon Geological Survey, Whitehorse, 1979.

———. "Ray Gulch Tungsten Skarn Deposit Dublin Gulch Area, Central Yukon" in Morin, James, ed., *Proceedings of Mineral Deposits of Northern Cordillera Symposium*, 1983: 245–54.

Mayo Historical Society. Gold and Galena: A History of the Mayo District. Mayo, Yukon: Mayo Historical Society, 1990.

MacLean, T.A. *Lode Mining in Yukon: An Investigation of Quartz Deposits in the Klondike Division*. Ottawa: Government Printing Office, 1914.

McKim, Major S.C. "Hundreds of Placer Miners Toiled in Vain in the Shadow of Marvelous Keno." *Dawson Daily News*, Mayo Edition, November 29, 1920: 28.

Nordin, Gary D. "1980 Assessment Report on the Dublin Gulch Property." Report prepared by BEMA Industries Ltd. for Canada Tungsten Mining Corporation Ltd. On file with Yukon Geological Survey, Whitehorse, 1981. http://yma.gov.yk.ca/090790.pdf

Polk's 1923–24 Alaska-Yukon Gazetteer and Business Directory.

R.L. Polk & Co.'s 1915–16 Alaska-Yukon Gazetteer and Business Directory.

Quandt, David, Chriss Ekstrom and Klaus Triebel. "Technical Report for the Fort Knox Mine." Prepared for Kinross Gold Corporation and Fairbanks Gold Mining Incorporated. Toronto: 2008.

Rowland, John. *Slipping the Lines: Adventures around the World in Peace and War*. North Battleford, SK: Turner-Warwick Publications, 1993.

Smyth, Stephen. *The Yukon's Constitutional Foundations, Volume 1: The Yukon Chronology*. Whitehorse: Northern Directories, Ltd., 1991.

SRK Consulting Engineers and Scientists. "NI 43-101 Preliminary Assessment Dublin Gulch Property—Mar-Tungsten Zone Mayo District, Yukon Territory, Canada." SRK Project Number: 173203. Lakewood, CO: SRK Consulting Engineers and Scientists, 2008.

"Victoria Gold." *Canadian Business Journal* (June 2016). https://www.cbj.ca/brochures/2016/Jun/Victoria_Gold/files/assets/basic-html/index.html#15

Wardrop. "NI 43-101 Technical Report—Feasibility Study, Eagle Gold Project, Yukon." Report prepared for Victoria Gold Corp., 2012, effective April 5, 2013.

Yukon Environmental and Socioeconomic Assessment Board. "Eagle Gold Project Screening Report and Recommendation Project Assessment 2010-0267." Whitehorse, 2013.

Yukon Geological Survey's Integrated Data System. Occurrence Number: 106D 025. Occurrence Name: Dublin Gulch. http://data.geology.gov.yk.ca/Occurrence/13910

Acknowledgements

It's a long, hard road to write a book of history, a journey made much better by those who help you find the way. I owe my interest in Yukon history to a giant of a man—Alan Innes-Taylor, who opened the door to so many pathways for me when I was young. Nor would this particular journey have been possible if John McConnell and Tara Christie had not asked me to write the history of Dublin Gulch in the first place.

My trip was made more pleasant because of my fellow traveller, my wife Kathy, whose interest in history has always spurred me on, who proved to be a good sounding board for my ideas along the way. Before my manuscript ever reached the publisher, she had looked at it, made suggestions, corrected grammar and corrected redundancies and inconsistencies that really brought my thoughts and words together.

There are some wonderful repositories of knowledge here that aided my quest for the Dublin Gulch story. The Yukon Archives has an outstanding collection of documents, photographs, maps, etc., and is always the first place I turn to when I embark on a historical project. The folks at the Energy, Mines and Resources Library also proved their worth; Chelsea Jeffery in particular found some amazing information when I thought I had reached the bottom of the barrel. Many old newspapers are now accessible on the internet, and when they are not, there are the microfilms at the Yukon Archives. The *Dawson Daily News* proved to be an invaluable source of information. The Historic Sites Unit of the Department of Tourism and Culture opened their files to me, for which I am thankful. Lowell and Lyn Bleiler are far more knowledgeable about the Mayo district than I will ever be, and I thank them for taking the time to point me in the right direction, and reviewing some of my early drafts.

The oral accounts of a number of people helped to put the flesh on the human side of the story. These people are Janet Dickson, Gordon Gutrath, Ron and Helen Holway, Sally Howson, Simon Mervyn, Frank Taylor and Howard White. Others who shared their personal connections to Dublin Gulch are Mark Ayranto, Keith Byram,

Michelle Dawson-Beattie, Tony George, Paul Gray, Sean Harvey, Daryn Leas, Brian Lennan, Letha MacLachlan, Kevin Mather, Ranj Pillai, Marty Rendall, Debra Ryan, Joe Sparling and Wendy Tayler.

Victoria Gold facilitated several visits to Dublin Gulch during the construction of the mine, from the shovel-turning in 2017 to the first gold pour in September of 2019. Tony George took me on a personal tour of the mine as it reached completion. Paul Gray took me on a tour of the Victoria Gold Nugget exploration camp in July 2019.

I know that behind the scenes, Lenora Hobbis was making many things happen for the various tours of the mine that I enjoyed. She also brought together hundreds of photographs of Dublin Gulch. I tracked down historical photographs, but she assembled hundreds of contemporary images, many of which are included in this volume. Among those whose photographs are contained in this book are Hugh Coyle, Mike Gunn, Paul Gray, Doug Hadfield (Bighouse Productions), Cathie Archbould, the Yukon Archives, and Lowell and Lyn Bleiler. Angus McIntyre, Brian Lennan and Frank Taylor kindly gave permission to use historical photos for this volume.

Several people were generous with their time in providing me with background information and suggestions for me to follow, including Lee Pigage, Bob Holmes and Anne Leckie.

Harbour Publishing has an excellent team with whom I have always enjoyed working. In particular, I would like to single out Silas White, who has transformed my rough draft from an uncut stone to a brilliant diamond. Peter Robson slaved over hundreds of photos to select those that made it into the book. What a daunting task. Nicola Goshulak edited the layout copy and added polish to the final product.

To all of these people, and any others who have been inadvertently left out, I express my thanks. Without them, it wouldn't have happened.

Image Credits

Page positions are indicated as follows: t=top, b=bottom, l=left, r=right, c=centre

Abbreviations

ALM	Alistair Maitland Photography
AM	Angus McIntyre Collection
BC	Bleiler Collection
BL	Brian Lennan
CA	Cathie Archbould Photography
CJ	Clive Johnson
DDN	*Dawson Daily News*
DH	BigHouseProductions.ca
GC	Government of Canada
TC	Taylor Collection
TT	Ted Takacs
VGS	Victoria Gold Staff
YA	Yukon Archives

Page	Source
iv–v	(Map) VGS
vi–vii	(Map) VGS
xx	Victoria Gold Staff
6	Shutterstock/optimarc
9	GC—MacLean, 1914 plate No. 29
12	DDN—December 22, 1910, p.17
14–15	YA—Mayo Historical Society Collection, M. Rich collection, 2012/19, *Gold and Galena*, p. 41
18	BC
21	YA—Mayo Historical Society Collection, Fred Taylor Collection, 2012/19, *Gold and Galena*, p. 40
22	TC
24	YA—Mayo Historical Society Collection, Fred Taylor Collection, 2012/19, p. 40

Page	Source
25	TC
27	AM—MHC-22
28tl	AM—HC-12
28tr	AM—MHC-13
28b	AM—MHC-18
29	AM—HC-2
30t	AM—HC-8
30b	YA—Mayo Historical Society Collection, A. Aho Collection, 2012/19, *Gold and Galena*, p. 39
31	BC
32–33	AM—HC-4
34	TC
35	BC
38–39	BL
40	TT
41	TT
43	TT
44	Shutterstock/Pavel Gaja
45	CJ
51	GC—MacLean, 1914 plate No. 27
52	GC—MacLean, 1914 plate No. 25
53	GC—MacLean, 1914 plate No. 4
54	GC—MacLean, 1914 Map No. 2
60	YA—Bill Hare fonds, 82/418, #6987
61	YA—Mayo Historical Society Collection, A. Aho Collection, 2012/19, *Gold and Galena*, p. 38
63	BL
67	VGS
68	VGS

Page	Source	Page	Source
69	VGS	121t, b	ALM
70t, c	From Hydrogeology Characterization and Assessment Report—New Millennium Mining Ltd. Prepared by GeoViro Engineering Ltd.	122	VGS
		123	ALM
		127	CA
		130–31	DH
70b	VGS	132	DH
71	VGS	133	Unknown
74	CA	134	Raine Mihtla
75	CA	135t, b	DH
76	CA	136t, b	DH
77	ALM	137	DH
78	VGS	138–39	DH
80	CA	140t, b	DH
81	CA	141t, b	DH
82t, b	VGS	142–43	DH
83	VGS	144	DH
84tl, tr	VGS	145t, b	DH
84b	CA	146–47	DH
86	VGS	149	DH
87tl, tr, b	VGS	150–51	DH
88	Rita Taylor/Banff Centre	152l	DH
90	VGS	152–53	DH
92	CA	154–55	DH
93	CA	156–57	DH
94	ALM	158l	DH
96	CA	158–59	DH
97tl	VGS	160tl, bl	DH
97bl, br	CA	160–61	DH
98	VGS	162	DH
100	VGS	163	DH
103	VGS	164–65	DH
105t, b	VGS	166	VGS
107t	VGS	167	DH
107b	VGS	169t, bl, br	CA
109	VGS	170	DH
110	VGS	172–73	DH
112	ALM	177	DH
114	DH	180t	VGS
116	CA	180b	VGS
118	DH	185	DH

Index

Note: Page numbers in **bold** *refer to images.*

Abbott, Nathan, 11, 17
Abbott, William, 11, 17, 58
Acheson, Jack, 26, 34, 36
Adams, Rod, 176
Aecon, 99
Agro, Hugh, 75, **76, 92,** 168
Aho, Aaro, 34, 62
Air North, 175, 181
Aitken, Thomas, 55
Alacer Gold Corporation, 99
Aldcroft, William, 48
Alkan Air, 181
Alverson, Jack, 55
Amax Gold, 66
AMEC, 112
Ames, R.S., 56
Arychuk, Kelly, 99
ATCO camps, 85
Aura Minerals Inc., 113
Avoca claim, 60
Ayranto, Mark, 72, 79, **80,** 83, 85,
    88–89, **98, 103, 107,** 113, 115–16,
    120, **121, 127,** 166, 168, **169,** 176,
    180

Bagnell, Larry, 120
Barker, Ed, 18–19, 23, 26, **31,** 34,
    61–62
Barrett quartz claim, 46
Barrick Gold, 102
Batty, Ruth, 62
BC/Yukon Chamber of Mines, 94
Bear Creek, 40
Bell, John Dease, 10

Belliveau, Joseph Edward, 10
Bema Gold, 75
Bema Industries, **45,** 65
Beringia, 2
Berry, William Morley, 46
BGC Engineering, 95
Big Springs property, 103
Binet, Gene, 10
Black, George, 12, 16, 25, 58
Black Creek, 11
Black Hills Creek, 94
Bleiler, Lowell, **34–35**
Bleiler, Theodore "Ted," 17, **18,**
    19, 43
Blodgett, V.V., 12
Blue Grouse claim, 49
Blue Lead group, 50, 52–53
BMC, 178
BNP Paribas, 111, 119, 122, 125, 128
Bob claim, 65
Bonanza Creek, 48
Bonnet Plume, 63
Bostock, Hugh, 11, 20, 23, 25
Bouvette, Louis, 57
Boyce, John, 62
Bralorne gold mine, 19
Bras d'Or Lime Company, 51
Bras d'Or Marble Company, 51
Breakwater Resources, 123
Brewery Creek mine, 91
Brimston, Sheriff, 13
Broadfoot, Kate "Kay," 23
Byng, Lord, 58
Byram, Keith, 65, 133, 174

Cairnes, Dr., 50, 56, 59
Camborne School of Mines, 112

Campbell, Robert, 3
Canada Tungsten (Can Tung),
    40, 42–43, 63–65, 117, 182
Canadian Expeditionary Force,
    16, 47, 56
Canamin Resources, 115
Canex Aerial Exploration Ltd., 63
Canex Placer, 63
Canfor, 81
Can Pro Development Ltd., 65
Cantin, Francois "Frank," 13,
    14–15
Cantin, Joseph, **14–15**
Cantin, Louis, 13, **14–15**
Cantin, Phileas, 13, **14–15**
Cantin Brothers, 13, **14–15,** 17–18,
    43
Carcross, 19, 171
Carmacks, 171
Carscallen, Frank, 48–49, 55, 185
Caterpillar Financial Services,
    128
Catto, Dr. William, 47–50, 59
Champagne, 175
Chateau Mayo Hotel, 34
Chilkat Tlingit people, 3
Christie, Dagmar, xii
Christie, James Murdoch "Jim,"
    16, 46–47
Christie, Jim, xii
Christie, Kara, xii
Christie, Sheamus, xii
Christie, Tara, xii, xvii, **94,** 95,
    124–25, 175, 181, 183
Coffee mine project, 178
Coleman, Corporal, 17
Colorado School of Mines, 93

Colt, J.J., 60–61
Commonwealth Bank, 111
Companies Act, 12
Congdon, Frederick, 58
Connaught Mines Ltd., 62
Corkery, James, 46
Corlin, Hugh, 46
Cornucopia Resources, 69
Council for Yukon Indians, 179
Cove McCoy property, 102–3
Coyle, Hugh, 79, 81, 168
Crisfield, George, 16, 46–47
Curniski, Orest, 42
Currier, H.A., 66
Curtis, Dan, 166
Cyprus Minerals, 66

Darron Placers Ltd., 36–37, 40
Dave claim, 65
Davidson, Jacob A. "Jake," 8, 10, 46, 55
Davis, John M., 46
Dawson City, 3, 8, 26, 43, 47, 52–53, 56, 58–59, 91, 94, 181
Dawson Daily News, 8, 12, 13, 49–50, 56–57, 59
Dawson-Beattie, Michelle, 175–76
Day, Clyde, 26
De Beers, 112, 123–24, 182
Denver Gold Forum, 166
Department of Indian Affairs, 4
Desarollos Minerals Ivanhoe Holdings Ltd., 66
Dickson, Gordon, 63–65
Dickson, Janet, 63–65
Discovery claim, 8, 10–11
Djukastein, Klaus, xv, 36–37
Dominion Creek, 26, 48
Dowling, Edward, 99
Downey, Patrick, 113
Drury, Bill, 26
Dublin Gulch: climate at, 22; derocker and sluice plant on, 38–39; early miners on, 14–15; history of, xvii, xviii; 1912 map of mining activity, 54; placer

mining at, 8, 12, 13, 20–23, 24, 34, 40, 42, 56–57, 68, 70, 78; population during mine construction, 171; promotional images for, 12, 13; rock derrick on, 21; transportation to, 11, 19, 23, 25, 43, 52, 59, 60, 61–62, 101, 104, 105
Dublin Gulch Mining Ltd., 42
Dublin Hydraulics Ltd., 12, 13
Dublin King claim, 48
Duensing, Darrell, 36–37, 40, 42, 63–64
Duncan Creek, 10, 36
Dye, Keith, 42

Eagle deposit or zone, 66–67, 71, 75, 76, 82–83, 86, 116–17, 124
Eagle Gold Mine: aerial view of site, 154–55, 172–73; blasting in production holes, 158; claims, 65; community impacts, 176–78; construction, 120–63; conveyors and heap leach facility, 162; crushers, heap leach facility and gold recovery plant, 160–61; drill rig in the open pit, 156–57; emergency response, 177; Emergency Response Team, 177; employees during construction, 171; employees during operation, 174; employment opportunities, 175–76; environmental impacts, 166, 176; Every Student, Every Day program, 177, 181; first gold pour, 166, 167, 168, 169–70; fleet of trucks, 114; heap leach facility, 164–65; impact on Yukon territory, 174–75; legacy for the Yukon, 178–82; lifespan estimates, 128, 183; on-site assay lab, 163; opening, 163; safety record, 178; shovel and haul truck, 160; shovel-turning ceremony, 121; sign displayed on gold recovery

plant, 149; substation and backup generators, 152; women in workforce, 17, 160, 163
Eagle Gold Project: aerial view of camp, 74, 84; agreement with First Nation, 99; auger drilling, 100; camp food, 137; capital for, 101–3, 110–11, 119; community consultation, 87; construction of crushers and support walls, 134, 136; conveyor system, 141–43; crusher facilities and conveyor system, 144; embankment for heap leach facility, 132; environmental assessment, 86; expanded camp, 109, 110, 122, 129, 130–31; exploration and permitting, 83; exploration program, 96–97, 117; feasibility study, 95, 108; full construction, 129; gold recovery plant, 146–47; haul roads and open pit, 118, 150–51; haul trucks, 135; heap leach facility and gold recovery plant, 145; Issued for Construction drawings, 133; major construction phase, 111; mining fleet, 114, 115, 128; pre-feasibility study, 92–93; safety record, 178; screen decks and crusher building, 140; "shovel-ready," 104; shovel-turning ceremony, 119–20; staffing, 148; water use licence, 106, 107, 134
Eagle group, 52–53, 56, 64
Eagle Pup valley, 70
East Potato Hills zone, 111
Edzerza, Don, 37
Electrum Strategic Opportunities Fund, 110
Elsa silver camp, 19, 184
Emerald mine, 93
Erickson, Mrs., 17
Evans, Mark, 55
Ewing Transport, 105

Fairbanks, 66
Fairbanks Gold Mining Inc., 66
Falcon zone, 111
Faro mine, 174–75
Ferrell, Helen, 47, 56
Ferrell, Jim, 47, 56
Fields Creek, 11
15 Pup, 42
Finning Canada, 113, 115, 128, 134, **135**
Finsch mine, 112
First Dynasty, 66, **67,** 69, 72, 86, 90
First Nation of Na-Cho Nyäk Dun: access and exploration agreement, 102; acknowledgement of contribution, 168; Comprehensive Cooperation Benefits Agreement, **98,** 99, 176, 179, **180;** conditions under Department of Indian Affairs, 178–79; connection with the land, 3; economic and employment opportunities, 163, 174, 177; environmental concerns, 91, 102; land claim settlement agreement and self-government, 179; negotiations and relationship with Victoria Gold, xiii, xv, 86, 88, **90,** 92, 120, 125, 129, 178–79; traditional territory, xi
First Nations: education opportunities, 181–82; employment opportunities, 175; impact of European colonization on, 4; land claim settlement, 179
Firth, T.A., 13
Fisher, Robert "Bobbie," 17, 19, 22, 47–50, 56–57, 59–60, 62, 185
Fisher Gulch, 42
Fort Knox gold mine, 65–66, 76, **90,** 91, 116, 183
Fort Reliance, 3
Fort Selkirk, 58
Fortymile, 4

Foundation claim, 48
Frank, Jack, 42
Friedland, Robert, 66
Fruta del Norte mine, 113

Galena Creek, 55
Galena Hill, 17
Galipeau, René, 123
Gateway Gold, 76–77
Gatey, John, 23
Gatey, Peter, 23
Geological Survey of Canada, 11, 50–51, 56, **60,** 61
George, Tony, 111, **112,** 120
Gibson, Jim, 18
Gill Gulch, **40–41,** 42
Gilmore Gold, 66
Golden Hill Ventures, 176
Gordon, J.K., 46
Gordon Landing, 10
Graham, George, 56
Grant, Vic, 17
Gray, Paul, 79, 81, 115, **116,** 117, 120, 183–84, **185**
Great Depression, 19
Great Plains Development Company of Canada Ltd., 62–63
Greaves, "Tiny," 17
Greer, Sheila, 3
Greig, Clifford, 23, 26
Guggenheim family, 53
Gustafson family, 8, 10
Gutrath, Gordon, 40, 42–43, 63
Gwich'in people, 3

H-6000 Holdings/Amax Gold, 65
Haddock, James, 11
Haddon, Timothy, 66
Hager, Chief Robert, 91
Haggart, Peter, 8
Haggart, Thomas "Tom," 8, 46
Haggart Creek, 8, 10–11, 13, **14–15,** 18–20, 23, 25–26, **27–33,** 34, 36, 40, 42–43, 58, **60, 78**
Haldane Bridge, 104, **105**
Hall, Annie, 50
Hamilton, Colin, 10

Hän people, 3
Happy Jack claim, 48, 53, 55
Harper, Arthur, 3
Harrington, Jim, **107**
Harvey, Sean, 75–76, **92, 127,** 128, 168
Hawthorne, Jack, 18
Henderson, Robert, 59
Heney, Thomas Edward, 46
Henson, Cyril Vernon, 46
Hiatt, Warren, 8
Highet Creek, 34
Hill, Christopher, 99
Hobbis, Lenora, xii
Holway, Helen, 37
Holway, Ron, 36–37, 40, 42–43, 63–64
Homer, Dr. Geoffrey, 23
Howson, Sally, 88–91, **107,** 120, **121,** 168, 179
Huckleberry copper mine, 115
Hudson's Bay Company, 3
Huffman, Grant, 55

Iditarod region, 53
Imperial Munitions Board, 16
Imperial Order Daughters of the Empire (IODE), 25
Indian River, 94
Iron Ore Company, 112
Ivanhoe Capital Corporation, 66
Ivanhoe Goldfields Ltd., 65–66

Jackson, Thomas, 17
Jackson's Saloon, 17
Janet Lake, 10
JDS Energy and Mining Inc., 108, 129, **133,** 178
JDS-Hatch, 113
Jeff claim, 65
Johnson, George, 66
Johnson, Harry, 42
Johnston, Thomas Nelson, 46
Johnston Creek, 46
Jones, Dr. C.M., 47
Jones, G., 46

Kappes, Cassiday & Associates, 95, 106

Karowe diamond mine, 113

Kazinsky, Louis, 18

Keele, Joseph, 11, 46, 56

Keno (steamer), 19

Keno City, 17, 19, 23, 57–58, 93, 184

Keno Hill, 17, 50, 57–58

Keno Hill Mining Company, 57, 60. *See also* United Keno Hill Mines Ltd.

Keobke, Neil, 61

King Edward claim, 48

Kinross Gold Corporation, 66, 75–76, 79

Kinsey, Agnes Jane, 48

Kinsey, Olive Powers, 48

Kirchner, Timo, **107**

Klondike district, 4, 7, 34, 46, 48, 59

Klondike Placer Miners' Association, 94

Kluane district, 40

Kluane Drilling Ltd., 108, 177, 184

Korean War, 26

Krol, Len, **92**

Kudz Ze Kayah project, 178

Lansing Post, 19, 47, 56

Lapierre House, 3

Leas, Daryn, 179

Leckie, Anne, 88

Lennan, Brian, 42, 64

Lightning Creek, 17

Lillooet, 19

Lochore, Norval, 17–19

Lone Star mine, 48–49, 53

Look-out Cabin, 52, **53**

Lunde, Vilhelm "Ole," 23, 25–26

Lundin Group, 113

Lynx Creek, 20

Lynx Dome zone, 111

M3 Engineering, 104

Mackenzie, George P., 10

Mackenzie, Thora, 10

Mackenzie King, William Lyon, 58

Mackenzie River, 46

MacLachlan, Dugald, **51**, 53, 88–89

MacLachlan, Letha, **88**, 89, 186

MacLean, Thomas Archibald, 51–53

MacLean party, **52**

MacLean report, **54**

Macquarie Bank, 111

Major, Arthur, 17

Malicky, Walter, 34

Margaret claim, 48

Mariposa Creek, 94

Martin, Archie, 17–18, 20, 60

Mar-Tungsten deposit or zone, **63**, 64–65, 67, 72

Mather, Kevin, **133**, 178

Mayo (formerly Mayo Landing), xii, xiii, 10–11, 13, 19–20, 22–23, 25–26, 47, 53, 56–58, 86, **87**, 88, 91, 93, **105**, 171, 175, 179, **180**, 181

Mayo, Al, 7

Mayo Creek, 7

Mayo district, 1–3, 8, 16, 46, 48–49, 52, 55, 57–59

Mayo Lake, 7

Mayo River, 7

Mayo Silver Mines Ltd., 62

McConnell, John, xi–xii, xiv, 75–76, **77**, 79, 81, 83, 85, **92–93**, 94–95, 102–4, **107**, 113, 117, 119–20, **121**, 122–23, 125–28, 163, **166**, 168, 175–76, 179, 181–83

McConnell, R.G., 50

McConnell glaciation, 2–3

McInnis, Mike, **92**

McIntosh, Alan, 10

McKay, Tom, 20, 60

McKim, S.C., 46

McKnight, Bob, 81

McQuesten, Jack, 3

McQuesten Lake, 184

McQuesten River, 3, 7–8, 23, 25, 40, 46, 55, 57, **60**, 61

McQuesten River Bridge, 176

McQuesten Substation, **153**

McWhorter, Harry, 46–47, 53, 55

Mervyn, Chief Simon, xi, xiii, xv, 37, 88, 92, **98**, 99, 102, 120, **121**, 129, 178–79, **180**, 182

Mexican claim, 49

Midas claim, 49, 55

Mill Canyon property, 102

Miner's Daughter, 166

Minto Bridge, 10–11, 52, 56

Minto mine, 113

Montague roadhouse, 19

Moskelund, J.E., 53

Mount Haldane, 52, **153**

Mount Nansen, 63

MRG Copper LLC, 103

Mullins, Ron, 37

Munroe, Curley, 49–50

Murphy's Pup, 42

Na-Cho Nyäk Dun Business Trust, 177

Na-Cho Nyäk Dun Development Corporation, 129, 176–77

Nanisivik mine, 93, 123

Nelson, Thomas, 7–8

Nelson Creek, 7–8

New Millennium Mining Ltd., 67, 69, 72, 86, 89, 91, 117

Newcomen, Warren, **107**

Newmont Mining Corporation, 72, 79, 101, 178

Nicol, Alex, 10

NND Cobalt Mine Services, 176

Norman Wells, 25

North Star quartz claim, 46

Northern Commercial Company, 17

Northern Tutchone people, 2–3

North-West Mounted Police, 10

Nugget camp, 183–84, **185**

Nugget zone, 111, 128

Nuway Crushing, 176

O'Brien, John, 56

Olive claim or zone, 48, 66, 86, 103, 106

Olive federal Crown grant, 72

Olive group, 52–53, 60
Olive Gulch, 57, 65
Olive mine, 18, 49–50, **51,** 62, 88
Olive-Shamrock zone, 108, 111
O'Loane, Dick, 17
O'Neill, John B., 60–62
Ophir claim, 49, 55
Orion Mine Finance, 123–26, 128
Ortell, George, 46
Osisko Gold Royalties Ltd., 123–24, 128

P. Burns & Co., 58
Patterson, Duncan, 10
Patterson, Frank, 91
Pelly, 58, 133
Pelly Construction, 65, 129, 174
Pershing Gold Corporation, 101
Peso Silver Mines Ltd., 61
Pillai, Ranj, 175, 178
Pitts, George, 58
Platinum Pup, 42
Plut, Frank, 42
Portlock, William, 17
Potato Hills, 2, 18, 47–48, **63,** 67, 106, 184
Potter, George, 20
Premier Gold Mines Ltd., 102
Premier Mine, 89
Prospectors and Developers Association of Canada, 94
Provencher, Conrad, 62

Queenstake Resources Ltd., 40, 63–65

R&D claims, 65
Ramey, Rod, 42
Raney, Martin Joseph, 48
Rawlins, Charles, 11
Ray, Elsie, 23
Ray, Harvey, 61
Ray, Irvin, 23, 25
Ray Gulch, 62, **63,** 64
Reed Creek, 40
Reid glaciation, 2
Relief Canyon property, 101

Rendall, Marty, 79, 85, **107,** 113, 120, **121, 123,** 124–28, 168
Rescan, 112
Rex-Peso zone, 111
Richards, T.C., 57
Rinfret, Raoul, 10
Rogue River, 46
Roots, Charlie, 2
Ross Creek, 11
Rouleau, Dave, 166, **169**
Roulette claim, 57
Royal Canadian Mounted Police, 17
Royal School of Mines, 112
Rupe, William, 7
Ryan, Debra, 181

Sadie-Friendship vein, 57
Savage, Alan, 115–16
Schellinger, A.K., 57, 60
Scroggie Creek, 94
Seaholm, Alberta, 23
Seaholm, Hugo, 23, 25
Securities Exchange, 81
Selwyn Basin, 1
Selwyn deposit, 72
Selwyn Mountains, 46
Shamrock zone, 66, 86, **110,** 128
Shandro, Jack, 23
Sharman, Victor, 42
Sheldon, Smoky, 64
Silver, Sandy, xiii–xiv, 120, **121,** 129, 166, **167,** 168, **169**
Silver Trail, **153**
Sime, Billy, 17
Sixtymile, 4, 94
Skagway, 19
Smashnuck, George, 26
Smit, Hans, 89–90
Smith, A.W.H., 12
Smoky claims, 64–67
Snap Lake mine, 124
Société Générale, 111
Sore Leg claim, 49
South Canol region, 89
Southern Tutchone people, 3
Sparling, Jim, **73**

Sparling, Joe, 175
Sprague, Bowles Colgate, 49–50, 53, 56, 59, 185
Stantec, 92, 95
Starmin Mining, 66
Stein, Steve, 90
Steiner zone, 66, 86, 111
Sterlite Gold Ltd., 72
Stewart, Jack, 47–50, 53, 56–57, 59
Stewart, Jessie, 17
Stewart and Catto group, 52–53, 55–56, 59, 185
Stewart gulch, 65
Stewart Pup, 53
Stewart River, 3–4, 7, 10, 13, 19, 46–47, 53
StrataGold Corporation, **70,** 72, **73,** 76–77, 79, 81, **82,** 83, 86, 93, 116, 124, 182
Strathcona Mineral Services, 93
Stride Exploration and Development Company, 62
Sugiyama, Jimmy, 17
Summit Camps, 129, **137**
Sun Valley Gold, 110
Sunnydale, 50
Silver King claim and mine, 55, 57
Suttle Gulch, 42
Suttles, John Jackson "Jack," 8, **9,** 13, 16, 43
Swanson, Mrs. Robert, 23
Swanson, Robert, 23, 25
Swede Creek, 42

Takacs, Ted, **40–41,** 42, **43**
Takara Resources Inc., 79
Tam Mines Ltd., 62
Tanana district, 66
Tayler, Wendy, 181
Taylor, Ann, 21–23, **25,** 26
Taylor, Bonnie, 36
Taylor, Frank, 22, **25,** 26, **34–35,** 36
Taylor, Fred W., 19–20, **21–22,** **24–25,** 26, 34, **35,** 36–37, 43

Taylor, Jim, 34, **35,** 36
Taylor, Joe, 66
Taylor, Joyce, 36
Taylor, Troy, 36
Taylor and Drury, 17
Tetra Tech, 95
Thompson, Dr. Alfred, 16, 56
Thompson, Dr. W.E., 12–13
Tin Dome, 62
Tinis, Scott, **107**
Tintina Trench, 2, 22
Tip Top claim, 61
Tourist Services (supermarket),
    34
Townsend, Norton, 17
Treadwell Yukon Company Ltd.,
    34, 57, 60
Trombley, Lloyd, 56
Trudeau, Pierre, 179
TSX Venture Exchange, 124, 168
Tucker, Terry, 72, **73,** 81
Turner, Jack, 10

United Keno Hill Mines Ltd.,
    62, 91

Van Bibber, Alex, 175
Van Bibber, Sue, 175
Victor Diamond Project, 112
Victoria claim, 47–48, 53, 55
Victoria Gold: access and
    exploration agreement, 102;
    acquisition of StrataGold, 77,
    79, 83; agreement with Yukon

Energy, 122; board of direc-
    tors, **88, 92–93,** 113, 126–27;
    community liaison, 91, 94;
    Comprehensive Cooperation
    Benefits Agreement, **98,** 99,
    176, 179, **180;** exploration, 86;
    financing and fundraising,
    xvii, 101, 110–11, 119, 122–29,
    168; legacy for the Yukon, xiv,
    178–82; negotiations and rela-
    tionship with First Nation, xv,
    90, 179; officers, xi, 72, **80–81,**
    111, **112,** 115, **116, 123, 166, 169;**
    Power Purchase Agreement,
    95; staff, xii
Victoria Gold Yukon Student
    Encouragement Society, 181
Victoria Gulch, 48
Victoria Resources, 75–76, 77,
    124. *See also* Victoria Gold
Vidette (steamer), 52

Waddco Placers Ltd., 26, 34, 36
Walby, William, 56
Wardrop Engineering Inc., 72,
    95, 101
Watson Lake, 63
Watters, Kit, 18
Wernecke Camp, 34
Westmin, 89
White, Heather, 110
White, Howard, 34, 36
Whitehorse, xii, 25, 94, 166, **167,**
    168, **169,** 174

Whitehorse Star, 163
Wilbur, Steve, 168
Wilkinson, Sonja, **160**
Williams, Chad, 75, 77, 79, **81,** 83,
    92, 93
Williams, William, 47
Williamson, Henrietta, 48
Williamson, W., 46
Wilson, Maynard, 26, 34
World War I, 16, 47, 56, 59
World War II, 22, 25–26, **60,**
    61–62

Yamasaki, Harry, 17
Yukon Consolidated Gold Cor-
    poration, 37
Yukon Energy Corporation, 93,
    95, 101, 122
Yukon Environmental and
    Socio-Economic Assessment
    Board, 95, 99, 102–3
Yukon Geological Survey, 94
Yukon Gold Company, 17, 57
Yukon Imagination Library, 181
Yukon Infantry Company, 16
Yukon Miners' Association, 50
Yukon Morning World, 11
Yukon River, 4, 8, 43, 53
Yukon Valley, 7
Yukon Water Board, **103, 107,** 134

About the Author

MIKE THOMAS

Michael Gates is the author of *From the Klondike to Berlin* (Harbour Publishing, 2017), which was short-listed for the Canadian Authors Fred Kerner Book Award. He is also the author of *Dalton's Gold Rush Trail: Exploring the Route of the Klondike Cattle Drives* (Harbour Publishing, 2012) and *History Hunting in the Yukon* (Harbour Publishing, 2010). He was formerly the curator of collections for Klondike National Historic Sites in Dawson City and pens the popular column "History Hunter" for the *Yukon News*. He lives in Whitehorse, YT.

Lost Moose is an imprint of Harbour Publishing Co. Ltd.
P.O. Box 219, Madeira Park, BC, VON 2H0
www.harbourpublishing.com

Front cover image and back cover inset by BigHouseProductions.ca
Back cover image Government of Canada, MacLean, 1914 plate No. 27
Project manager Peter A. Robson
Edited by Silas White
Indexed by Ellen Hawman
Cover and text design by Roger Handling/Terra Firma Digital Arts
Layout by Linda Gustafson/Counterpunch Inc.

Printed and bound in Canada

Harbour Publishing acknowledges the support of the Canada Council for the Arts, the Government of Canada, and the Province of British Columbia through the BC Arts Council.

Library and Archives Canada Cataloguing in Publication
Title: Dublin Gulch : a history of the Eagle Gold Mine / by Michael Gates.
Names: Gates, Michael (Historian), author.
Description: Includes bibliographical references and index.
Identifiers: Canadiana 20200211951 | ISBN 9781550179408 (hardcover)
Subjects: LCSH: Gold mines and mining—Yukon—History. | LCSH: Gold miners—Yukon—History.
Classification: LCC FC4011 .G37 2020 | DDC 971.9/1—dc23